ETHNOBOTANY
OF THE
WAIMIRI ATROARI INDIANS
OF BRAZIL

THE ETHNOBOTANY OF THE WAIMIRI ATROARI INDIANS OF BRAZIL

WILLIAM MILLIKEN
ROBERT P. MILLER
SHARON R. POLLARD
ELISA V. WANDELLI

"We are the generation for whom the only message for a tropical biologist is: Set aside your random research and devote your life to activities that will bring the world to understand that tropical nature is an integral part of human life. If our generation does not do it, it won't be there for the next".

D.H.Janzen, 1986

ROYAL BOTANIC GARDENS
KEW

First published 1992

Addresses of authors:

William Milliken, Royal Botanic Gardens, Kew, Richmond, Surrey TW9 3AE, U.K.
Robert P. Miller, Programa Waimiri Atroari, Rua Sucupiras 310, Conjunto Kissia, 69.040 – Manaus – AM, Brazil.
Sharon R. Pollard, Freshwater Research Unit, Zoology Department, University of Cape Town, 7700 Rondebosch, Cape, South Africa.
Elisa V. Wandelli, Universidade do Amazonas, Departamento de Biologia, Estrada do Contorno 3000, 63.060 – Manaus – AM, Brazil.

General Editor: J.M. Lock

Special Editors for this Volume: D.V. Field & J.A. Ratter

Cover Design by Media Resources, RBG, Kew.

ISBN 0 947643 50 8

Typeset at Royal Botanic Gardens, Kew by Pam Arnold, Christine Beard, Margaret Newman, Pam Rosen and Helen Ward.

Printed & Bound in Great Britain by Whitstable Litho, Whitstable, Kent.

This book is dedicated to the Waimiri Atroari Indians

CONTENTS

ABSTRACT

The Waimiri Atroari are a Carib-speaking tribe of Amazonian Indians, whose population has declined since intensive contact with the outside world commenced in the 1970s. A study of their ethnobotany was carried out in 1989 on behalf of Programa Waimiri Atroari. Their history and their current situation are discussed. Quantitative data are presented for their use of the 214 tree and liana species in one hectare of terra firme forest, to 79% of which useful properties were attributed. Seven principal plant use categories: food, technology, medicine, construction, fuel, ritual and commerce are discussed in the light of the recent changes which the tribe has undergone. Plant-use data are presented for over 300 species, together with information regarding their distribution and ecology, and references to previous records of their useful properties. The importance of the forest to the Waimiri Atroari is examined.

RESUMO

Os Waimiri Atroari são uma tribo amazônica do grupo linguístico Karib, cuja população tem sido muito reduzida desde contatos intensivos com a civilização branca na década de 1970. Realizou-se um estudo da sua etnobotânica em 1989 para o Programa Waimiri Atroari. Sua história e situação atual são abordadas. São apresentados dados quantitativos sobre a utilização de 214 espécies de árvores e cipós encontrados em um hectare de floresta de terra firme, das quais 79% foram atribuídas propriedades úteis. Estabeleceu-se sete categorias principais de uso das plantas: alimento, tecnologia, remédio, construção, combustível, rituais e comércio; as quais foram relacionadas às recentes mudanças sofridas pela tribo. Dados sobre o uso são apresentados para mais de 300 espécies de plantas, juntamente com informações referentes à distribuição e ecologia, e outros registros de propriedades úteis. A importância da floresta para os Waimiri Atroari é examinada.

INTRODUCTION

For a long time it has been recognized that Amazonian Indians possess an outstanding knowledge of the plants and animals of their natural environment, and that prior to their contact with the outside world they were entirely reliant upon those resources for their existence. The remarkable extent of that knowledge, however, and the degree to which those resources are utilized, are only just becoming evident in the light of recent studies (e.g. Prance *et al.*, 1987). This realisation coincides with a growing public awareness that the Amazonian rainforests, like those of most of the rest of the world, are disappearing at such an alarming rate that their demise no longer lies in the distant future, but is a matter for immediate concern.

Sadly, economic pressures appear to demand financially viable arguments for the preservation of these forests. Whether or not their future could realistically be secured by sustainable utilization of their resources, and whether such projects could be implemented without further exploitation of their inhabitants, is currently the subject of substantial debate (e.g. Peters *et al.*, 1989; Gray, 1990; Posey, 1990a; Wright, 1991). If this is to be the case, however, a thorough knowledge of the sustainable economic potential of the forests' specific components, and of how mankind can maintain and profit from their existence, is a prerequisite (Posey, 1983). That knowledge rests with the indigenous inhabitants of the forests, for whom thousands of years of experience and experimentation have left an inheritance of information whose value is inestimable*. The acculturation and extinction of Amazonian tribes, a process of long standing and one which has by no means ceased, has already caused the irretrievable loss of the majority of that information, and deprived Amazonia of great cultural wealth. Furthermore the removal of these peoples from their traditional lands has deprived the forests of their only experienced tenants and guardians, and opened the way for their destruction.

Traditionally, ethnobotanical studies have been of a purely qualitative nature. The first estimate of the percentage of useful trees in a plot of land in Brazilian Amazonia was made by Carneiro (1978), with the Carib speaking Kuikuru tribe of the upper Xingu River. According to his survey results, 76% of the trees in the plot were recognized as useful (Prance *et al.*, 1987). Between 1983 and 1985, similar studies, this time supported by herbarium voucher specimens, were made with four other Amazonian tribes: the Ka'apor and Tembé of Brazil (Balée, 1986; 1987); the Chácobo of Bolivia (Boom, 1987); and the Panare of Venezuela (Boom, 1990). The combined results of these studies are presented in Prance *et al.* (1987), and reveal tree usage percentages of 77, 61, 79 and 49 respectively for a hectare of *terra firme* (unflooded) forest. Prance included in his list neither firewood species nor those which provide food for game animals, and if these two attributes are included (as they were in the original publications of Balée), then the tree utilization figures for the Ka'apor and Tembé both become 100%.

Such high levels of utilization of forest tree species are by no means restricted to unadulterated indigenous groups, however. Recent studies of a 7.6ha

*In considering the actual (monetary) value of this knowledge, it is essential that the complex issue of intellectual property rights is confronted and clarified (see Posey, 1990b).

communal forest reserve in northeast Peru, run by 'ribereño' communities of detribalized Indians and 'mestizos' (people of mixed blood), demonstrated a 60% proportion of potentially useful tree species (Pinedo-Vasquez *et al.*, 1991).

Stable, peaceful contact with the Waimiri Atroari has only been established since the 1970s, and is a phenomenon not unrelated to the intimidatory influence of the large armed force which occupied their territory during that period. Nevertheless their customs, beliefs and lifestyle were, at the time that the present study was carried out, substantially modified from their original state. This process of modification will to some degree have commenced as soon as the tribe entered contact with the outside world, but subsequent to their 'pacification' it accelerated dramatically. Recent events have stimulated an unusually heavy presence of FUNAI (Indian Foundation) officials in the Waimiri Atroari reserve, a body whose stated policy involves the integration of Indians into the national society. As a gesture of compensation for the violation of their territory, a 25-year programme of 'support' (Programa Waimiri Atroari) has been implemented which necessitates the presence of teaching, medical and logistic support personnel throughout the reserve.

An influx of this kind, regardless of good intentions or potential benefits, will inevitably have rapid and radical influences on the Indians, and the dangers of irreversible cultural damage are vastly magnified by an initial ignorance of the workings of that culture. Previous data on the Waimiri Atroari are scanty. In the 1880s João Barbosa Rodrigues, Director of the Jardim Botânico de Rio de Janeiro, made several expeditions to the Indians then known as Crichaná. His anthropological and ethnobotanical observations (Barbosa Rodrigues, 1885) differ in certain aspects from those of the present study, and leave room for doubt as to whether the tribe with whom he lived were really the group known today as the Waimiri Atroari. In addition to some notable ethnographical discrepancies, many of the most basic words recorded in his Crichaná vocabulary bear little or no relation to those currently spoken by the Waimiri Atroari (see p. 13).

In the early part of the 20th century, the Waimiri Atroari were visited briefly by the Austrian anthropologist Richard Payer and by the German Theodor Koch-Grünberg. Subsequently, until as late as the 1980s, contact was largely limited to the 'sertanistas' (frontiersmen) of the Indian Protection Service (SPI/FUNAI). Padre Egydio Schwade of the Roman Catholic Church came to work with the tribe in 1985, but was expelled from the area by FUNAI in 1986. Marcio Silva, a linguist/anthropologist from the University of Campinas (UNICAMP), conducted research over several months during 1987 until he too was expelled. The most lengthy and detailed studies to date were conducted by Stephen Baines, an anthropologist from the University of Brasília, who spent 17 months between 1982 and 1985 investigating the effects of FUNAI's contact methodology and psychology upon the Waimiri Atroari (Baines, 1988). Baines has since been disallowed from returning to continue his research (Pereira, 1990).

The only recent ethnobotanical study of these Indians was carried out by Pierre Grenand in 1987, with a team of postgraduate students from the Amazonian Research Institute (INPA). This survey entailed a rapid examination of 0.1 hectare of forest, carried out along the lines of Boom and Balée's studies. The estimated potential use of the trees in the plot was 65% (Miller *et al.*, 1989). These results, however, are not directly comparable with those of Boom and

Balée, nor with those of the present study, since the minimum diameter of the trees examined by them was lower (6cm as opposed to 10cm). Furthermore, instead of consisting of uniform *terra firme* forest, their study site encompassed a variety of forest types in order to maximize diversity.

For the sake of comparability, the present study, conducted between August and December 1989, duplicates the quantitative hectare study methods employed by Balée and Boom, and has been presented in a format similar to that used by Vickers & Plowman (1984) for their study of the Siona/Sequoia, and by Boom (1987) for the Chácobo.

The aims of the study were as follows:

To determine, quantitatively, the degree to which the Waimiri Atroari utilize their forest resources.

To record part of the Indians' knowledge of their useful forest plants
— a permanent archive, available to the Indians themselves.
— as a source of information for the Programa Waimiri Atroari.

To determine the effects of increased external contact upon the Indians' relationship with their environment.

It should be stressed that this work does **not** purport to be the definitive account of Waimiri Atroari ethnobotany. Although additional information has been included, it is primarily a survey of their use of the trees in one hectare of *terra firme* forest. These represent but a fraction of the plant resources available in their reserve, and many years of combined botanical and anthropological research would be necessary to document the subject thoroughly.

THE WAIMIRI ATROARI

History

The Waimiri Atroari are a forest tribe of Indians of the Carib language family. Traditionally they inhabited vast tracts of the land north of the Rio Negro and east of the Rio Branco, an area of which they were probably the sole occupants once the Aruak Indians had moved north to Guiana in the 1780s (Barbosa Rodrigues, 1885). Subsequently external influences provoked them to retreat to the smaller, less accessible rivers, and at present they are restricted to a range of approximately 2.5 million hectares to the north of Manaus (Fig. 1). This territory theoretically includes the entire basin of the Rio Alalaú (a tributary of the Jauaperí), and parts of the basins of the Rios Camanaú, Curiuaú and Uatumã.

Prolonged contact with Luso-Brazilians did not occur until the mid 19th century, when hunting, rubber-tapping and Brazil-nut collecting communities began to penetrate the Waimiri Atroari lands from settlements on the banks of the Rio Negro. Such contact almost invariably resulted in bloodshed. The Waimiri Atroari, who have defended their lands against invaders and aggressors with perhaps more tenacity than any other Amazonian tribe (Hemming, 1987), attacked the settlements of Airão, Pedreira and Moura, and temporarily drove the inhabitants from the latter. In reprisal, military expeditions were sent up the Jauaperí where appalling massacres were perpetrated. In 1874, for example, Lt Antônio de Oliveira Horta, acting under the instructions of the Governor of Amazonas State, spent an entire year 'hunting' Indians on the Jauaperí river (Barbosa Rogrigues, 1885).

One incident from that episode, related by Barbosa Rodrigues, amply demonstrates the contemporary attitude towards the Indians. On the 21st of November a band of Indians were chased into a lake, where many of them were shot. 'Many died; a few escaped and hid in the forest.' 'The next day, in order that the inhabitants of Moura should feel secure, he returned with troops to the scene of the previous day's bloody drama. There he found 23 wounded Indians, all hiding silently amongst the foliage of the trees which they had climbed. Then began a scene of barbarism. They were like enthusiastic huntsmen before a troop of howler monkeys. Everyone wanted his part in the slaughter. They pointed their guns, fired, and a poor Indian would fall amongst general hilarity. Thus they all came down, but for one who was caught up on a branch.'

Not altogether surprisingly, such conflicts created a mutual atmosphere of distrust and aggression that remained unbroken for many years. The first significant peaceful contact was made in 1884 by the botanist and anthropologist João Barbosa Rodrigues, who had been sent to the area by the Brazilian government. He attempted a reconciliation between the Indians and the settlers, with some degree of success, but upon his departure the old enmities resurfaced and with them the bloodshed. By the start of the twentieth century the Indians were friendly towards visitors who approached them peacefully, but a retaliatory massacre in 1905, ordered by Governor Constantino Nery of Amazonas, inflamed the situation once more. On that occasion, in response to the alleged ousting of a rubber tapper from his house, 283 Indians were killed. Many had fled to their houses to avoid capture, and were burned alive within them

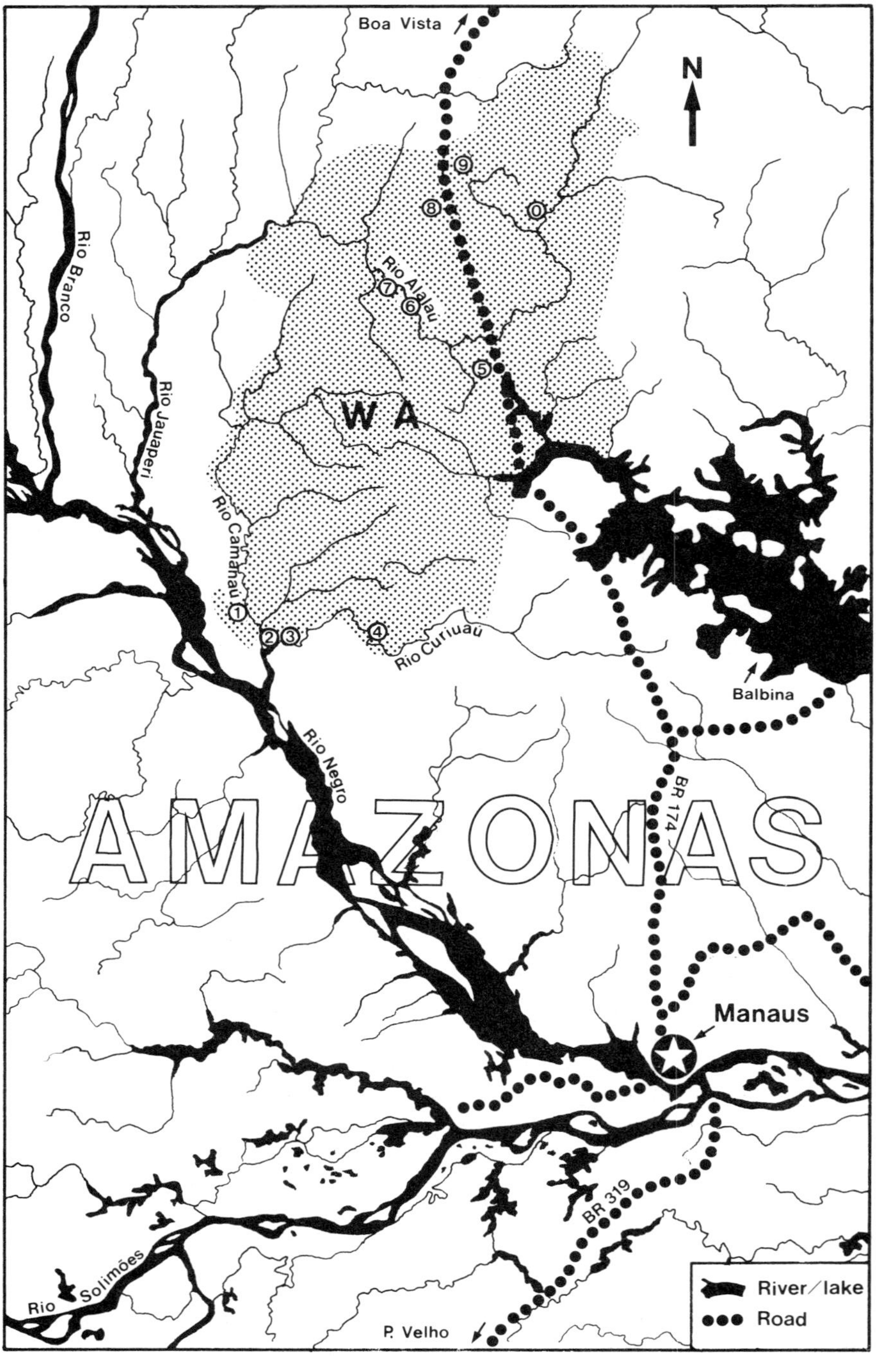

Figure 1. Location map (scale 1:2,500,000) of the Waimiri Atroari reserve (stippled area). The boundaries are approximate, as are the positions of the ten villages (1 Maré, 2 Cacau, 3 Curiaú, 4 Sumauma, 5 Mynawa, 6 Jara, 7 Alalaú, 8 Jawara, 9 Xeri, 0 Maikon).

Figure 2. Location of Fig. 1 within South America.

(Hemming, 1987). The Waimiri Atroari were understandably enraged, and relations with outsiders reverted to their original bloody state.

In the early 1970s the BR 174 (Fig. 1), the highway linking Manaus to Boa Vista (Roraima State) was finally pushed through Waimiri Atroari land. The construction of this road marked, for the Waimiri Atroari, the beginning of a period of rapidly escalating contact with the outside world. In an attempt to 'pacify' the Indians prior to the commencement of construction, a team led by Padre João Calleri had been sent up the Rio Sta Antônio de Abonari in 1968. Calleri's expedition was received amicably by the Indians, but unfortunately his attitude towards them was somewhat primitive, as he demonstrated in his radio messages to Manaus. When, for example, the Indians showed enthusiasm for relieving the expedition of its 'merchandise', he stated: 'nothing was permitted (the Indian well knows that this is our right as superior beings)'. He is also said to have threatened a man with a shotgun who did not want to leave his (Calleri's) hammock (Carvalho, 1982). The result of Calleri's insensitivity was, predictably, that he and all but one of his party were killed. Road construction went ahead, implemented by the Sixth Engineering and Construction Battalion of the Brazilian army (BEC), and the inevitable conflict ensued.

FUNAI 'sertanistas' (frontiersmen) had by that time made considerable progress in achieving friendly contact with the Waimiri Atroari, and had set up contact posts in several locations. They were led by Gilberto Pinto Figueiredo, a man who exacted high standards of professional behaviour from his men and who had become a close friend of Maruaga, one of the charismatic leaders of the tribe (Plate 1). The army's presence within the reserve was a cause of major setbacks to FUNAI's contact attempts. Posts were attacked and FUNAI employees killed, and in December 1974, as Gilberto attempted to curb the aggressive behaviour of the military construction team and allay the breakdown in communications, he was mysteriously killed (Carvalho, 1982; FUNAI, 1975).

Gilberto's replacement, Sebastião Amâncio da Costa, was a man of somewhat different temperament. He was quoted in the newspaper 'O Globo' (6/1/75) as having said: 'The Waimiri Atroari need a lesson: to learn that they have done something wrong. I shall use an iron hand against them. The chiefs will be punished and, if possible, taken far from their lands and people' (MAREWA, 1983). He claimed to be sick of FUNAI's 'war without weapons', and maintained that the traditional pacification strategy of FUNAI had failed (General Rondon's motto for FUNAI's predecessor, the ill-fated SPI, was 'Die if necessary, kill never'). He decided that it was time to use more direct means such as dynamite, grenades, tear gas and submachine guns to give the Indians 'a demonstration of the force of our civilization' (Davis, 1977).

At the same time that the road was being built, preliminary studies were under way for the construction of Balbina, a massive hydroelectric dam on the Rio Uatumã (Fig. 1). On 25 February 1986, by which time the construction of the dam was virtually completed, this project was authorized by presidential decree. The dam had been built for the Eletronorte electricity company, and was intended to furnish the city of Manaus with all of its electrical requirements. Input of water into the lake, however, is insufficient to provide the desired output of power*. The dam was principally financed with World Bank funds, and is generally accepted today as having been an unmitigated disaster (Fearnside, 1990). Approximately 250,000 hectares of rain forest were inundated, including a sizeable tract of the Waimiri Atroari reserve. The inhabitants of two Waimiri Atroari villages were displaced (Plate 6a) and are now receiving compensation in the form of Programa Waimiri Atroari: a 25-year programme of educational, medical and material support which began with the essential demarcation of the boundaries of their reserve

The Waimiri Atroari territory was originally designated the title 'Indigenous Reserve' on 13 June 1971. In 1979, however, the Timbó mining company (subsidiary of the Paranapanema mining group) received prospecting rights for cassiterite (tin ore) within its limits. In 1980, representatives of FUNAI and the anthropologist Hildegart Hick made a rapid overflight of the north-eastern extremities of the reserve — the area of interest to Paranapanema — and alleged that no Indians existed within the region. Although shortly after this another FUNAI anthropologist, Angela Baptista, stated that the reserve *should definitely* include those disputed north-eastern territories, President Figueiredo nonetheless dismantled the reserve in 1981. He revoked the three decrees that had created it, and set aside in its place a smaller 'temporarily prohibited area for

* One of the more outrageous proposals for the subsequent supplementation of this input is the diversion of the neighbouring Rio Alalaú (Gribel, 1990).

pacification and attraction of the Waimiri Atroari Indians'. This prohibited area excluded the 526,800 hectares of the Waimiri Atroari land that were being prospected — and subsequently exploited — by Paranapanema (Plate 6c). This somewhat complex and controversial episode is elaborated upon further by Baines (1991) and Porantim (1990).

During this period the Waimiri Atroari were undergoing a process of intensive acculturation, implemented by the FUNAI workers operating within their territories. Maré (Plate 3c), the village at which the present study was carried out, was no exception in this respect. In 1980 nearly all of the survivors of the epidemics which had ravaged the villages on the Rio Camanaú and its tributaries were brought down to live at the old FUNAI post near the mouth of the river, and in 1981 they were transferred to Maré. There they lived alongside over 20 non-Waimiri Atroari people, principally FUNAI employees (and their families) with limited education, no anthropological guidance or training and little understanding of cultural values different from their own. In 1983 a group of the Indians living at Maré decided to return to their old malocas (villages) upstream, but were ordered back to Maré by FUNAI. They were subjected to a rigid labour schedule, working in gangs on the large plantations that had been set up on their behalf (Plate 3e), and receiving payment in the form of meals at the FUNAI post. They were forced to wear clothes, and were mocked by the FUNAI workers who publicly derided their traditional beliefs and customs. The Waimiri Atroari, who were at this point seriously culturally disorientated, began to reject much of their past and to assimilate the attitudes and ideas of the people amongst them (Baines, 1988).

In 1982 officials of FUNAI signed a contract (Contrato 039/82, since recognized as illegal) with Paranapanema, authorizing the construction of an access road to the Pitinga mine through Waimiri Atroari land. The study group that was set up in 1985 to investigate this contract were all dismissed prior to completing the job. Finally in 1987, in spite of the formulation of further contracts which assigned the mineral rights of the entire area to the Paranapanema mining company, that area was redesignated as a reserve and was demarcated.

The decree that recreated the reserve guaranteed to 'defend the entire hydrological system of the Rio Alalaú (Plate 3b), bearing in mind the presence of cassiterite exploration that could have ecological effects on the tributaries of that river'. In contravention of this, FUNAI officials then made further contracts with Paranapanema (Termo de Compromisso 001/88 and Termo de Re-Ratificação 001/88), allowing them to retain certain tributaries on the left bank of the Alalaú in return for financial and logistic compensation such as the implantation of cattle-ranching (see p. 119). As a consequence of these (unconstitutional) contracts, the Alalaú has since been undergoing significant pollution by the mine's effluents (Plate 6d) which enter via the Igarapé Tiarajú: one of the tributaries left in the hands of Paranapanema. Fish stocks are said to have been affected, and Programa Waimiri Atroari medical personnel have reported an increase in the incidence of gastro-enteritis in the Indians living downstream.

Representatives of Paranapanema, in a meeting held with the 'captains' of the affected villages in 1989, claimed that nothing could be done to prevent the pollution of the Alalaú, but offered to install drinking wells and fish farms in compensation. Wisely this offer was turned down, and a legal solution is currently being sought (via the Brazilian government) on behalf of the Indians.

People

Calleri estimated the Waimiri Atroari population to be around 3000 in 1968, although earlier estimates were as high as 6000. It is hard to tell to what degree the ensuing decline was due to the atrocities reported to have been committed during the construction of the BR174 (bombing of villages, electrocution of Indians on barbed-wire fences, vaccine overdoses etc.), and how much was due to the disease epidemics which simultaneously ravaged the tribe, but by the mid-1980's their numbers were down to around 330. By 1989, when this study was carried out, it had climbed again to over 450 and was continuing to recover, although it has since been suggested that the rate at which it is doing so is decreasing (Silva, 1991).

It is often stated that the Waimiri Atroari were originally two separate tribes, whose unification was a reaction to their shared conflict with an alien and threatening culture. In fact it seems likely that although there were (and are) dialect differences between their component sub-groups, they actually constituted a single ethnic entity. This is a complex issue which probably merits further investigation. Those Indians remaining today, who are (as are all Amazonian peoples) the few survivors from what was once an extensive network of villages, have now agglomerated in the 10 settlements which were set up in convenient locations by FUNAI and Eletronorte within the reserve. Intermarriage is common between these villages, but each maintains its identity as either Waimiri or Atroari, in spite of the fact that the division was created arbitrarily by SPI/FUNAI as part of their 'indigenist' policy, and subsequently accepted by the Indians themselves (Baines, 1988).

The villages traditionally consist of a single maloca or round-house (Plate 3c), in which the 50 or so inhabitants' hammocks are distributed in family groups around the periphery. The majority of the villages are now located on the main rivers, some on 'black earth' sites which have clearly been favoured by Indians for many generations. They have little contact with other Indian tribes, although they do maintain a small trade with the Waiwai (also of the Carib linguistic family) whose territories lie to the northeast. According to Barbosa Rodrigues (1885), the Crichaná used to engage in trade with the Tarumã Indians, who originally inhabited part of the present Waimiri Atroari territory but merged with the Waiwai at the end of the 19th century (Grenand, pers. comm.). Nowadays goods are generally exchanged for bead bracelets and necklaces, which the Waimiri Atroari do not produce themselves. There is one group in close proximity, the Pirititi (Piruichichi), who inhabit the head-waters of the Rio Branquinho and have yet to establish contact with Brazilian society. The Waimiri Atroari, who recognize the precarious position of their reduced population, claimed to be keen to liaise with these Indians. But on at least one past occasion when contact was made between the two groups, the Pirititi exhibited no reciprocal desire for friendship and as a consequence were attacked and killed.

Plate 1. Old photograph of Waimiri Atroari men wearing traditional aroid-root waist bands (no longer used). The second man from the left is the leader Maruaga. To his right (wearing a hat) is Gilberto Pinto Figueiredo, who was killed in December 1974 (see p. 8). The young man crouching in the foreground is Bixkia Taira Pedro, who contributed to the information collected during the present study. Photo courtesy of FUNAI (Manaus).

The main staple of the Waimiri Atroari diet is manioc (cassava), which is grown in swidden clearings in the forest. Work is communal: the felling and burning is the responsibility of the men, the planting is undertaken by both men and women together, and the harvesting is carried out by women only. Produce is shared amongst the members of the village, as are the proceeds of the hunt. Hunting is generally effected with bow and arrow, and is strictly the preserve of the men. A few shotguns have recently been introduced and these are now favoured for the killing of large game (see Yost & Kelley, 1983), although ammunition supplies are irregular. Fish form a large and important part of the diet.

As a 'trekking' people, the Waimiri Atroari would traditionally have moved their villages when available game became scarce. Imposed sedentariness, however, has created the need to travel further from the village to hunt than would originally have been necessary. Major hunting expeditions, involving most or all of the village menfolk and lasting for several days, are necessary to provide sufficient food for the *marba*, backbone of Waimiri Atroari culture. *Marba* is a continuous 3-day ceremony of dancing, singing and eating where every aspect of the Indians' lives and beliefs is reiterated in song, where friendships and marriages are made and broken, children are initiated and symbolic rituals performed whose meanings are as yet little understood. These ceremonies, which are held at apparently irregular intervals, generally involve several villages and help to maintain the cohesion of the tribe. They took place during the study period, but were not observed in detail.

The Crichaná

The paucity of previous information regarding the customs and lifestyle of the Waimiri Atroari inevitably hinders any analysis of the effects of contact upon their use of plants. The earliest of such records were made by João Barbosa Rodrigues, but there remains a shadow of doubt as to whether his observations were made with the same people. According to his records (Barbosa Rodrigues, 1885), the correct name for the Indians with whom he made contact was Crichaná, and he maintained that the name Waimiry (which was sometimes applied to them) was in fact a Portuguese corruption of Jauapery, the name of the river on which they lived. The name Waimiri, however, also holds a significance in contemporary *lingua geral* (*wai* = arrow, *miri* = small), and the arrows of this tribe are indeed short in comparison with those of the other tribes of the Guiana plateau (Grenand, pers. comm.). His descriptions of their large round houses with two doors, their head-dresses for ceremonies, the woven fibres around their wrists (see Plate 4b), their disgust for tobacco, their weapons, and the notable proportion of myopic and albino individuals amongst them conform closely with observations made during the present study*. Similarly his descriptions of Crichaná attire — the 'aprons' of seeds worn by the women and the belts of araceous aerial roots worn by the men — sound remarkably similar to those used by the Waimiri Atroari until a few years ago (see Plate 1).

*There are also individuals with apparently 'mestizo' (mixed blood) characters among the Waimiri Atroari. This may be due in part to their reported association with Brazilian fugitives in the last century (cf. Carvalho, 1982), or to later liaisons.

There are, however, some notable differences which indicate either that the group studied by Barbosa Rodrigues was a different one from that with whom the present study was made (although closely related), or that in the intervening years considerable behavioural changes have taken place within the tribe. He noted, for instance, that the Crichaná all pierced their ears and threaded them with short lengths of knotted string, that they often painted their faces red, and that they sometimes wore false beards of howler monkey hair. At their ceremonies they wore special necklaces of wood and toucan feathers, and feather decorations on their arms, and drummed thick columns of bamboo on the ground to provide an accompanying rhythm for their dances. Superficial linguistic comparisons (based on the vocabulary given by Barbosa Rodrigues (1885)) do little to elucidate the situation. Certain extremely basic words (e.g. 'fire') are virtually identical to those used today, whereas other words for equally mundane objects (e.g. 'water') bear no similarities whatsoever. It has been suggested that these differences could perhaps be attributed to the Macuxi Indian interpreters employed in those early investigations (Baines, pers. comm.).

Certain practices observed by Barbosa Rodrigues, such as the use of *Bertholletia excelsa* and *Lecythis* fruits as food/drink vessels and the ribs of turtles as fruit graters, could understandably have been lost as a result of the advent of new technology. But the disappearance of the traditional practices mentioned in the last paragraph, none of which appear to be extant within the Waimiri Atroari today, would indicate rather more deep-seated cultural change (not inconsistent with the policies of acculturation which SPI and FUNAI have practised with these Indians over the last sixty years). One of the more surprising differences is that according to Barbosa Rodrigues' records the Crichaná would wrap their dead in their hammocks and inter them, standing up, in the hollow of a living tree. But according to Carvalho (1982), the Waimiri Atroari burn their dead and throw the ashes into the river (at least until recently), and this was supported by the note made during the present study that the wood of a *Ferdinandusa* species (Rubiaceae) was commonly used for cremations in the past.

DESCRIPTION OF THE STUDY AREA

The village of Maré (Plate 3c) is located approximately 200 km NW of Manaus on the right bank of the Rio Camanaú, at an altitude of about 60m. At the time of study, the total population of the village was around sixty. The Camanaú is a black-water tributary of the Rio Negro, which joins the Rio Solimões to form the Amazon proper shortly downstream of Manaus. Maré lies on the approximate geological boundary between the gneisses, granites etc. of the 'Guiana' complex and the sandstones and conglomerates of the 'Prosperança' formation (RADAM, 1978). The soils in the region are classified as yellow clay latosols (RADAM, 1978), and relief is minimal. The two nearest meteorological stations are Manaus and Barcelos, the former having an average annual rainfall of approximately 2100mm and a mean annual temperature of 26.7°C, and the latter of approximately 2000mm and 26°C respectively (Villa Nova *et al.*, 1976). There is a reasonably marked dry season between the months of June and October, which is more accentuated in Manaus (downstream of the Camanaú) than it is in Barcelos (upstream).

The predominant vegetation type in the area is evergreen *terra firme* (unflooded) forest (Plate 3a), although there are patches of 'campina' and 'campinarana' (low forest types on sandy soils) present, and a stretch of 'igapó' (seasonally flooded) forest along the margins of the Camanaú River. In the hectare of forest sampled in the present study (61°15'W, 1°45'S), 200 species of trees were collected (with diameter at breast height (dbh) > 10 cm) representing 32 families. There were 645 individuals in the plot, with a total basal area of approximately 31m^2. The nine dominant families, ranked in order of Importance Value Index (a measure of dominance, density and frequency), were Leguminosae (*sens. lat.*), Burseraceae, Sapotaceae, Moraceae, Chrysobalanaceae, Lauraceae, Meliaceae, Myristicaceae and Palmae. Those species with more than ten individuals in the hectare were *Clathrotropis macrocarpa* (Leguminosae), *Eschweilera coriacea* (Lecythidaceae), *Protium hebetatum* (Burseraceae), *P. apiculatum*, *Oenocarpus bacaba* (Palmae), *Eschweilera grandiflora*, *Swartzia ulei* (Leguminosae) and *Eschweilera rhododendrifolia*. A more detailed analysis and description of this forest will shortly be published elsewhere.

FIELD METHODS

The principal technique employed in this study was a quantitative tree utilization survey of one hectare of *terra firme* forest. Preliminary visits to the area indicated the region of the Camanaú/Curiuaú rivers (Fig. 1) as being the most suitable for study, since the Indians in that area are the most isolated and consequently their culture and traditions would seem likely to be the most intact. Given, however, the recent history of treatment of the Indians on those rivers (see p. 9), this may not actually have been the case.

Access to the study area was obtained by boat. The village of Maré, the furthest up the Rio Camanaú, was selected as a study site and a hectare plot of *terra firme* forest, in the form of a 500m x 20m N-S transect, was set up. In order to minimize the likelihood of previous disturbance, a site was chosen approximately 15 minutes walk from the village itself. In the immediate vicinity of the village a considerable number of trees had been felled in the past for construction, or occasionally for fruit gathering. Care was taken to find an area representative of the local *terra firme* forest, avoiding the damp depressions with their differing floristic composition.

The transect was divided into 25 sub-plots of 20m x 20m, and within each sub-plot every tree and vine with a diameter at breast height (1.3m) of 10cm or more was tagged with a numbered aluminium plaque secured with an aluminium nail. Specimens were collected from each tagged individual (with the exception of the bacaba palm *Oenocarpus bacaba*), and dried using standard botanical collecting techniques. A complete set of the voucher specimens is lodged in the herbarium at UTAM, (Instituto de Tecnologia da Amazônia, Manaus), and fertile material at Kew (K), Manaus (INPA) and in certain cases Bergen (BG), Edinburgh (E), New York (NY), St. Louis (MO), Tempe (ASU), Utrecht (U) and Washington (US). In addition, records were made of the trees' heights, trunk diameters, buttress root sizes (where present), and bark characteristics (including exudates).

Details of the useful properties of the trees, their Indian names, their relationships with animals, and any other significant information were collected in three phases. The principal informants were Kwaxi Manelão, Pawyna Elias, Warpeki Comprido, Bixkia Taira Pedro, Arie José, Krepi Ramiro and Aminia Camilo. These names, both Portuguese and Waimiri Atroari, are apparently not the Indians' true personal names but nicknames created at the request of FUNAI officials. The Waimiri Atroari believe that if they repeat their true names it can bring illness upon them (Baines, pers. comm.).

'Phase one' of data collection took place in the forest. Each tree was visited with one or more of the informants and the details were recorded. Since many of them spoke very limited Portuguese, it was generally necessary to enlist the help of younger members of the village for translation, or to use a tape recorder. Waimiri Atroari words and plant names were recorded throughout in the orthographical format which, during the period of study, was being taught to the Indians by the Programa Waimiri Atroari. However that format has since been modified, and the words presented here have been adjusted accordingly by Edilberto Macedo Fonseca (Programa Waimiri Atroari). Details of pronunciation are given in Appendix I.

'Phase two' of the information-gathering was conducted in the village (Plate 4a). The specimens were sorted into morphotypes, and one example of each was shown to a group of informants. This was carried out over a period of several days. The resulting data were then compared with those from 'phase one'. Where the names or given uses differed significantly between the two, the morphotype was referred to 'phase three'. Here a specimen was taken to its parent tree in the forest in the presence of a group of informants, and an attempt was made to clarify which, if either, of the sets of information collected in the first two phases was correct. This was not possible in every case.

In addition to the hectare study, miscellaneous plants which occurred outside the plot, and to which specific uses were attributed by the Indians, were collected. Efforts were also made to collect those plants whose products were noted in the village. Preliminary identification of the majority of the material was made in the INPA herbarium in Manaus. Most determinations were subsequently checked at Kew, and in certain cases specimens were sent for examination by specialists in the relevant taxonomic groups (see Acknowledgements, p. 120).

WAIMIRI ATROARI PLANT NOMENCLATURE

The Waimiri Atroari recognize a number of vegetational and geomorphological units within their reserve, of which the most important is *txi*, or *terra firme* forest. The word *txi* also means 'territory', and the fact that to these Indians their lands and their forests are synonymous offers a reflection of the significance of that habitat for these people. The *terra firme* forest is the primary resource of the Waimiri Atroari, and it was the chief object of this study to demonstrate the extent to which they utilize it. Nonetheless each of the other vegetational categories recognized, although spatially less significant, yields important plant products which are not found elsewhere. These include *iyweprikiapo* (seasonally flooded black-water 'igapó' forest), *amehe* and *dare'* (low 'caatinga' forest types on sandy soils), *maraha* (secondary 'capoeira' vegetation), *wypy* (highland), *krasa* (lake), *pa'* ('baixada' drainage course), *mixi paty* ('buritizal' *Mauritia* palm swamp) and *wesi* paty ('açaízal' *Euterpe* palm swamp).

Berlin *et al.* (1974), in their detailed study of Tzeltal plant nomenclature, recorded 116 unambiguous expressions for plant parts and described those people as 'astute ethnophytographers'. Without such a detailed analysis and such an intimate knowledge of the language, it would be dangerous to attempt too definitive a translation for the nomenclature of a subject as complex and closely linked to Waimiri Atroari culture as the forest. For example one complication which clouds the issue of direct and precise translation is the occurrence of polysemy, as exhibited (for example) by the word *wiwe*. This word, although used as a life-form label for tree, is polysemous with wood and with certain wood products (such as wooden arrow-heads), as is likewise the equivalent Ka'apor Indian word of *mira* (Balée, 1989b). However, a brief survey of Waimiri Atroari nomenclature of plant and forest anatomy was conducted, and the data are presented in Fig. 3 (in diagrammatic form to minimize the errors potentially caused by preconception).

It seems that the Waimiri Atroari use a comparatively simple nomenclatural system for plant anatomy, albeit a highly logical one. The name used for tree trunk, for example, is the same as that for leaf midrib, branch the same as secondary leaf venation, petiole as pedicel, etc. Their structural division of the vegetation into the three vertical sections *kawyny* (*kawy* = high), *kybyny* (*kyby* or *kypy* = close) and *kytany* (*kyta* = low) corresponds closely with that recorded by Posey (1987) for the Kayapó. Trees (*wiwe*) are sub-classified according to their uses or characteristics, e.g. *arawyta wykry* (howler monkey's food), *pyrwa iepry* (arrow tree), *xiriri ieptxi* (cold cure), *i'ri* (fruit tree) etc. A list of these words is presented in Appendix II.

The majority of the trees recorded in the study hectare were also assigned 'botanical' names. These names may be very specific, or they may encompass several species. Furthermore some species may be attributed alternative names, for example *Couma macrocarpa* (Apocynaceae) which is apparently referred to either as *kamnjeni* or *kamnjany*. The significance of these variations is as yet uncertain (in this particular case the words are very similar and the difference may be attributable to individual variation), but it is worth noting that Balée (1989b) has recorded similar synonymy for certain plant species among the Ka'apor Indians (whose language is Tupi-Guarani).

The degree of nomenclatural species differentiation appears to depend to a large extent upon the useful properties of the plants. Thus a tree whose characteristics or uses are unique or rare, or one which is regularly used (such as *Symphonia globulifera*, Guttiferae), will almost certainly be assigned a 'specific' name (i.e. referring only to that species). Berlin (1976), in his discussion of the concept of rank in Aguaruna botanical nomenclature, refers to these names as *monospecific generic names* (genus being the smallest class which does not require much close study to recognize), and these accounted for 82% of the generic names among the Aguaruna people. Groups of species with common or similar uses or characteristics may be grouped under broader generic names,

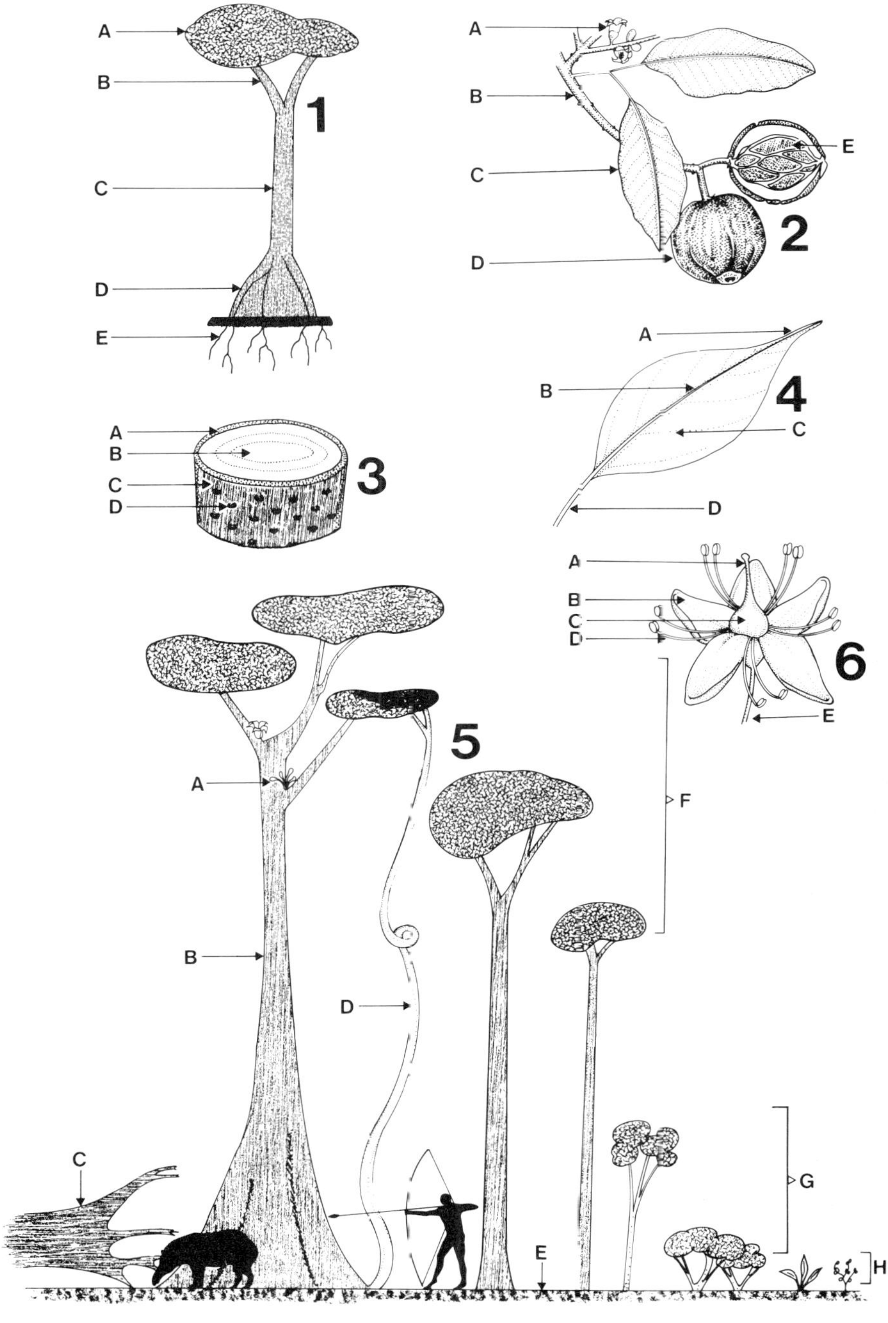

Figure 3. Waimiri Atroari nomenclature of plant and forest structure (schematic representation). 1) Tree: 1A) *ary*; 1B) *kata'*; 1C) *epy*; 1D) *tximtxa*; 1E) *nitxi*. 2) Flowering/fruiting branch: 2A) *epyryry*; 2B) *kata'*; 2C) *sowypia*; 2D) *ape*; 2E) *dia*. 3) Section of tree trunk: 3A) *bixi pyny*; 3B) *pyny*; 3C) *bixi*; 3D) *iararakyry*. 4) Leaf: 4A) *nisi*; 4B) *epy*; 4C) *ekba/kata'*; 4D) *dety*. 5) Forest: 5A) *wiwe pykyny*; 5B) *wiwe* (fruit-tree = *i'ri*); 5C) *iahyryry*; 5D) *wiwe pyky iykyny*; 5E) *sowypia iawytakyha*; 5F) *kawyny*; 5G) *kybyny*; 5H) *kytany*. 6) Flower: 6A) *nisimyty*; 6B) *demxi*; 6C) *apremy*; 6D) *xyhy*; 6E) *dety*. Modification of a drawing by Glenn Rodrigues.

which may or may not be further subdivided, depending again upon their significance. Plants designated the same Waimiri Atroari name in Table 3 will therefore tend to have the same uses and interspecific relationships attributed to them.

The allocation of names conforms very closely with our own familial or generic classifications. All of the Meliaceae, for example, were classified as *swakira,* almost all of the Lauraceae as either *penji* or *awa'ba,* and virtually all of the Burseraceae as either *wara'* or *aiywa.* In the cases of the Burseraceae and the Lauraceae there was considerable inconsistency in the choice of name given for a particular species, and it may be that the alternatives allocated are either interchangeable or depend upon some factor other than the tree's species (such as size).

It is not always easy to distinguish between a specific 'botanical' Waimiri Atroari name and one which is simply being applied to a plant to describe its features or uses. Hartmann (1967), in a study of Borôro plant nomenclature, does not attempt to distinguish between the two but lists all given names under 15 nomenclatural categories. In some cases it appears that for the Waimiri Atroari the two are synonymous, e.g. *kyme'* (= comb = *Apeiba echinata*). Perhaps because only one plant produces combs (the spiky fruits of *Apeiba),* there is no need to assign it a botanical name for differentiation from others. However, other plants may be assigned 'botanical' names (e.g. *amoty* = fibre = *Guatteria* spp.) which, although referring to a particular species or group of species, may also serve as usage denominators for other species with different 'botanical' names (see below).

Scientific name	W A usage denominator	W A 'botanical' name
Apeiba echinata	*kyme'*	*kyme'*
Guatteria olivacea	*amoty*	*amoty*
Sterculia pruriens	*amoty*	*tykexsowypry*

The Waimiri Atroari 'usage and characteristic denominators' recorded during this study are presented in Appendix II. 'Botanical' names are given in the overall list of plants. Although the majority have not been effectively translated, those which have been appear to fall into at least three broad categories:

1. Names comparing the plant with, or relating it to, an animal e.g. *wiamy iemry* = *Callichlamys latifolia* (Bignoniaceae) = 'tortoise's penis'; *iatwa iepry* = *Cecropia* spp. (Moraceae) = 'ant-tree'.
2. Names describing the plant's physical characteristics e.g. *minja iepkry* = *Protium fimbriatum* (Burseraceae) = 'tapioca-sap'; *xuia pyny kiriri* = *Abuta sandwithiana* (Menispermaceae) = 'yellow-wood vine'.
3. Names describing the plant's uses e.g. *mare iepry* = *Sorocea guilleminiana* (Moraceae) = 'sieve-tree'.

Many of the names recorded fall into the first of these categories; Hartmann (1967) recorded a similarly high incidence of the inclusion of animal names in Borôro plant nomenclature. This phenomenon is said to be characteristic of hunting peoples (Hartmann, 1967).

The provenance of a plant, whose 'botanical' name may not be confined to one species (as has been discussed), is sometimes used as an 'identifier' for specificity. A number of other criteria may be used for such differentiation, including leaf colour, size, myrmecophily etc. *Hevea* (Euphorbiaceae), for example, grows both in *terra firme* forest (*H. guianensis)* and in igapó forest (*H. spruceana),* but both are known as *kynyry ba.* They may be distinguished as *syna kany kynyry ba (Hevea* from the water) and *txi tany kynyry ba* (*Hevea* from the forest), respectively.

It is interesting to note that certain plant names appear to have been borrowed or adapted from the Portuguese (or *lingua geral)* vernacular names used in the region. In the case of those apparently borrowed from *lingua geral* (Tupi-Guarani), such as *tawyba* (*Mezilaurus* sp., Lauraceae) which is probably derived from 'ita-úba' (stone tree), this borrowing may have taken place a long time ago. Some plant names may also have been borrowed from non-Carib tribes which inhabited the region during the period of 'formation' of the modern tribe, such as Tarumã and Arawak. In some cases, however, the borrowing of plant names seems to have been in the reverse. The vernacular names 'pataua' and 'arouma', (*Jessenia bataua* and *Ischnosiphon arouma*), for example, seem likely to have been derived either from the Waimiri Atroari names *patwa* and *warma*, or if not then at least from another Carib language.

WAIMIRI ATROARI PLANT USE

The great majority of the plants collected during the present study were attributed useful properties by the Waimiri Atroari. These have been divided into six categories (Food, Technology, Medicine, Construction, Ritual and Commercial), each of which is discussed below:

Food Plants

The Waimiri Atroari are a hunting/horticultural trekking society, who practise small-scale swidden (slash-and-burn) agriculture in much the same way as do the majority of native (and settled) rural Amazonian communities. The agricultural systems of such Amazon peoples have in many cases been shown to be extremely complex and highly developed (e.g. Salick, 1989; Posey, 1984). Not only are their management systems complex but they also employ extraordinarily high numbers of cultivars of certain crop species. For example in a study of four villages of Tukano Indians in the northwest Brazilian Amazon, the inhabitants were found to recognize (by name) approximately 137 cultivars of manioc (cassava) amongst their swidden clearings, which represented at least 55 distinct forms (Chernela, 1987). Similarly, amongst the Kayapó Indians of Brazil, up to 12 cultivars of bananas have been recorded within a single clearing, and 22 cultivars of sweet potato (Kerr, 1987).

No detailed analysis of Waimiri Atroari agricultural methods has yet been carried out. To a large extent the complexity of their system will depend upon the degree to which horticulture forms a part of their traditional way of life (as opposed to that subsequent to contact), and that is extremely difficult to assess.

The three main crops grown are manioc (see Plate 5d), of which both 'sweet' and 'bitter' varieties are used, banana and sugar cane, which are cultivated in small clearings in the vicinity of the villages (see Fig. 4). Neither banana nor sugar cane are, however, native to Amazonia.

The forest is felled and burned towards the end of the dry season (September), in order to coordinate the subsequent planting with the onset of the rains. These clearings are used for several years, and are generally expanded annually until the available area of fertile land has been fully exploited. In the past this event would probably have provided one of the major stimuli for the subsequent translocation of a village. The Waimiri Atroari actually attributed the diminished productivity of an old clearing to the presence of malevolent spirits, which are themselves likely to have reinforced their desire to leave. A number of studies (e.g. Irvine, 1989; Balée & Gély, 1989; Denevan *et al.,* 1984) have demonstrated that many Amazonian tribes visit fallow swidden sites for several years after their abandonment, in order to collect fruits, medicinal plants etc. from the succession of pioneer species which recolonize the area. Furthermore these fallow sites tend to be richer in game species than the surrounding forest (Balée & Gély, 1989; Denevan *et al.,* 1984) and are used for hunting. Although the Waimiri Atroari have been noted to revisit these plots for the collection of fruits, arrow canes etc. from the previously planted crop species, their exploitation of the pioneer species has not been studied.

Originally the clearings made by the Waimiri Atroari were principally situated in the immediate surroundings of their houses, but in recent years there appears to have been an increasing tendency for the Indians to locate their swidden sites further afield. This may be a result of increased pressure on the nearby land arising from the reduced frequency of translocation of the villages, or it may be due to the increasing availability of mechanized transport. Also, the size of some of those clearings appears to have increased (Plate 3e), at least partly as a consequence of the efforts of FUNAI employees to increase the Indians' 'productivity' (see p. 45). Certainly the advent of the chain-saw will have made the clearing of forest a far less onerous task than it was in the past, and the recent introduction of cattle will likewise have had a very significant effect upon land use practices.

Between February and July 1990 a survey of the swidden sites of nine of the 10 Waimiri Atroari villages was carried out by Rodrigo de Azevedo Lima, then Coordinator of Programa Waimiri Atroari's 'Sub-Programa de Meio Ambiente e Apoio à Produção' (Azevedo Lima, 1990). The average clearing size was found to be approximately 4.9ha (some of the villages had more than one clearing), with a *per capita* swidden area of 0.17ha. The Indians at Maré maintained that the presence (or abundance) of certain plants in an area, such as a particular species of *Selaginella,* is indicative of suitability of the soil for cultivation. Whether these plants actually reflect soil fertility, or whether their presence is a result of previous human disturbance, is uncertain. However, high densities of *Selaginella* have elsewhere been shown to be frequently associated with base-rich soils (Ratter *et al.,* 1988).

All of the clearings investigated by Azevedo Lima were located on sites of previous occupation, attested by the presence of ceramic sherds in the soil. At least six identifiable ceramic 'phases' have been recorded in the soils of the lower Rio Negro region, and at many sites there may be two or more present

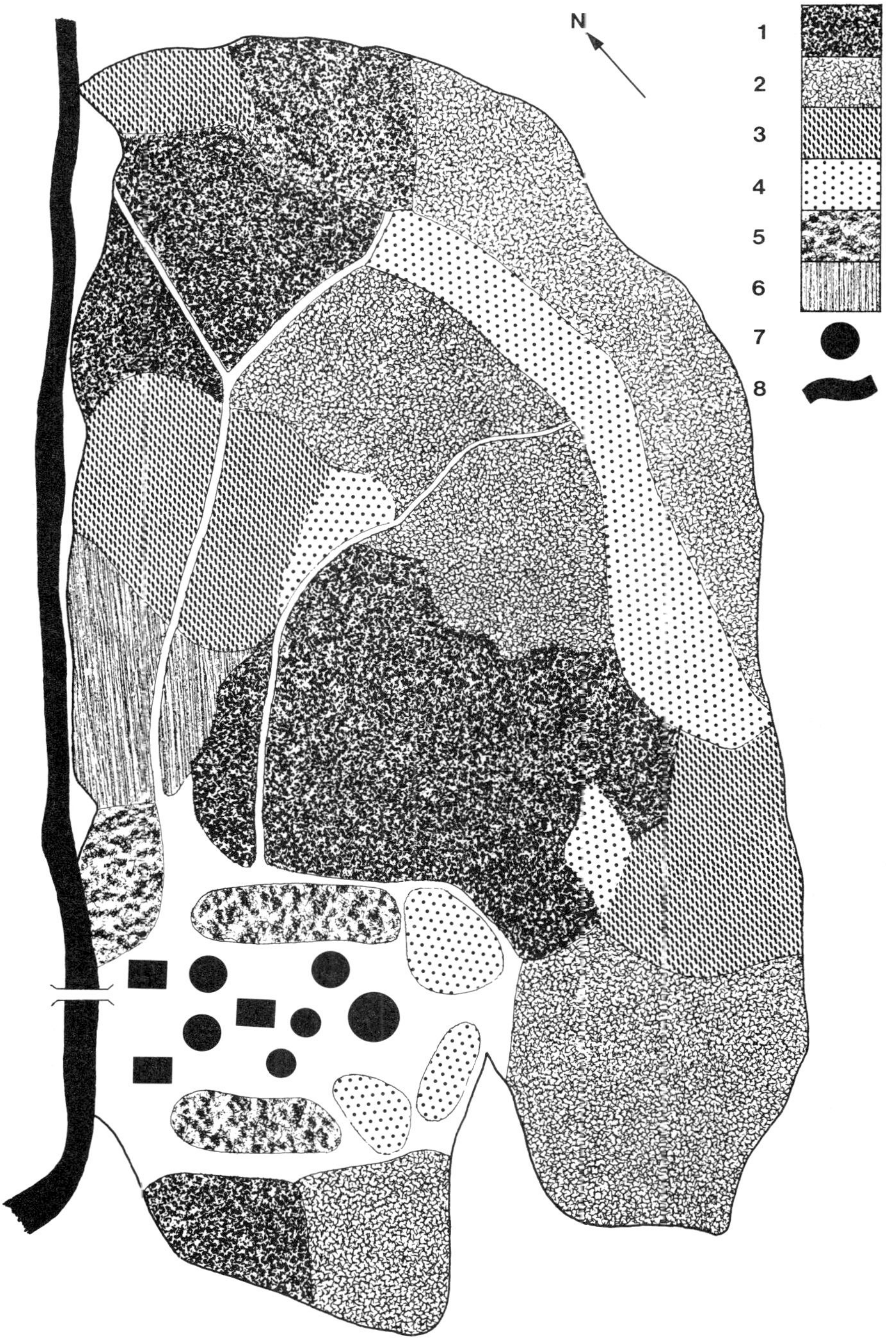

Figure 4. Swidden agriculture at the Waimiri Atroari village of Xeri. From an original drawing by Edilberto and Edith Fonseca (Programa Waimiri Atroari) in 1989: 1) manioc (bitter); 2) manioc (sweet); 3) sugar cane; 4) banana; 5) fruit trees; 6) arrow grass; 7) building; 8) stream. Areas in white represent forest, except around the buildings where the ground is open. The total area under cultivation is approximately 6.6ha.

(Simões, 1974). The preferential use of such anthropogenic 'terra preta' (black earth) soils for cultivation has been recorded amongst many Amazonian indigenous groups. These soils, which are common and widely distributed within the Amazon basin, contain high levels of phosphorus, calcium and carbon, and are thought to mark the sites of previous villages (Smith, 1980).

The Asurini and Arewete Indians of the Xingu basin, for example, inhabit ancient settlement sites where potsherds of the Tupi-Guarani, Tapajoara and Marajoara styles can be found (Balée, 1989a). From the depth of these soils it is apparent that the sites have been occupied, at least sporadically, for many hundreds of years. Some of the cultures represented in the archaeological record there have long since been extinct — indeed they are so long forgotten that the Arewete believe the sherds to be of divine origin (Balée, 1988). The Waimiri Atroari, who claimed that their ancestors never made pottery in some of the styles found in the soil beneath their villages, denied knowledge of their origins.

A list of the cultivated species encountered during the above survey is given in Table 1. In addition, although the present study was primarily concerned with the Indians' use of forest plants and no systematic study of the swidden systems was carried out (a separate study of cultivation systems is due to be undertaken shortly by E.V.W.), a list of the other species observed in cultivation at Maré village is provided. Many of the species, particularly the fruit trees which are grown around the houses, have been introduced very recently and are present in only one or a few of the villages. Others will have been introduced during earlier contact with agents of SPI/FUNAI, or traded with other indigenous groups. One cultivar of banana, for instance, is said to have been received in trade from the Waiwai Indians, and when the ill-fated Padre Calleri visited previously uncontacted Waimiri Atroari villages on the Rio Sta Antônio de Abonari in 1968, he was offered an orange!

According to legend, the ancient Waimiri Atroari ate only *kararba*, a type of palm (unidentified) which grew wild in the forest and provided an edible starch. One day an Indian came across a particularly large specimen, and although he tried to cut it down the effort was too great for him. Presently a piranha appeared from the river, and asked him whether he would like him to cut it down. Gratefully the Indian agreed, and went down to the river to bathe. When he returned the piranha had already felled the great tree, and advised him to cut its branches and to plant them in the ground. This the Indian did, and thus appeared manioc.

Today the Waimiri Atroari's carbohydrate requirements are almost entirely met by the products of their swidden cultivation, but nevertheless two alternative sources were identified in the forest: the seeds of *Clathrotropis macrocarpa* (Leguminosae) and the tubers of *Humirianthera rupestris* (Icacinaceae). Both of these species were recognized as potential 'famine foods'*, and were said to have been used as staples in the past. Both also contain poisonous substances which, prior to eating, must be removed from them by a similar process (grating and washing) to that used to extract the cyanogenic glycosides from manioc (*Manihot esculenta*).

* When food is scarce, the Waimiri Atroari claim to eat clay from the river banks. The eating of clay or earth has also been recorded amongst several tribes of the Guianas (Roth, 1924).

Table 1
Species cultivated by the Waimiri Atroari (Azevedo Lima, 1990)

Species	Waimiri Atroari name	Portuguese name	English name	Family
Anacardium occidentale	*hiri*	caju	cashew	Anacardiaceae
Ananas comosus	*nana*	abacaxi	pineapple	Bromeliaceae
Annona montana	*araxki*	araticum	wild soursop	Annonaceae
Annona muricata [2]	*araxki*	graviola	soursop	Annonaceae
Artocarpus altilis [1]	-	fruta pão	breadfruit	Moraceae
Bactris gasipaes	*merepy*	pupunha	peach palm	Palmae
Bertholletia excelsa	*tetky*	castanheira	Brazil nut	Lecythidaceae
Bixa orellana	*kiswe*	urucum	anatto	Bixaceae
Bromelia sp.	*karwa*	caraua	caraguata	Bromeliaceae
Capsicum sp(p).	*warasy*	pimenta	pimenta	Solanaceae
Carica papaya	*mabaia*	mamão	pawpaw	Caricaceae
Citrullus lanatus [1 3]	*marasia*	melançia	water melon	Cucurbitaceae
Citrus aurantifolia [1 2]	*arexa*	limão	lime	Rutaceae
Citrus sinensis [1 2]	*arexa*	laranja	orange	Rutaceae
Cocos nucifera [2]	*kyky*	coco de Bahia	coconut	Palmae
Crescentia cujete	*bixiwa*	cuieira	calabash	Bignoniaceae
Cucurbita sp.	*xirimu*	abóbora	pumpkin	Cucurbitaceae
Dioscorea trifida	*nepy*	cará	yam	Dioscoreaceae
Genipa sp.	-	jenipapo	genip	Rubiaceae
Gynerium sagittatum	*mapyry*	flecheira	arrow cane	Gramineae
Inga edulis	*akai*	ingá	ice cream bean	Leguminosae
Ipomoea batatas	*mepe*	batata doce	sweet potato	Convolvulaceae
Malpighia punicifolia [3]	-	acerola	Barbados cherry	Malpighiaceae
Mangifera indica [1]	*manga*	manga	mango	Anacardiaceae
Manihot esculenta	*minja*	mandioca	manioc	Euphorbiaceae
Musa sp. [1 2]	*wyky*	banana	banana/plantain	Musaceae
Psidium guajava	-	goiaba	guava	Myrtaceae
Rollinia mucosa	*briba*	biriba	queenfruit	Annonaceae
Saccharum officinarum [1 2]	*awaxe*	cana de acúcar	sugar cane	Gramineae
Syzigium malaccense [3]	-	jambo	rose apple	Myrtaceae
Theobroma grandiflorum [3]	*ak'a*	cupuaçu	patashte	Sterculiaceae

[1] = Species alien to the Neotropics
[2] = Species introduced prior to Programa Waimiri Atroari
[3] = Species introduced by Programa Waimiri Atroari

Clathrotropis macrocarpa was the dominant tree species in the hectare study plot (11% of all the trees recorded), and its seeds have been found to contain toxic quinolizidine alkaloids (G.P. Lewis, pers. comm.). Similar use of these seeds for food* has been reported from the Yanomami Indians of northern Brazil (Cocco, 1972), and the use of the tubers of the vine *Humirianthera rupestris* as a food source has also been recorded elsewhere in Amazonia. Richard Spruce noted its use amongst the Indians of the Rio Negro, and reported that 'when properly prepared, the farinha of (*H. rupestris*) is scarcely distinguishable from that of mandiocca'. Improperly prepared, however, it was seen to be highly toxic (Sandwith, 1956). This species is also used as a fish poison by the Waimiri Atroari (see p. 30).

According to a myth recounted (to E.V.W.) by an Indian of Mynawa village, the forest was so low in the days of the Waimiri Atroari ancestors that Indians wandering amongst the trees could pick the fruits from their branches. But the Waimiri Atroari abused these trees, stripping them of their fruits, and Mawa (the first Indian) punished them by transforming it to its present height. Nonetheless, fruits provide an important source of vitamins and minerals in the Waimiri Atroari diet, and although the diversity (and maybe quantity) of fruit trees cultivated around their villages (see Plate 3d) is currently increasing due to external influences, the forest remains a vital source. In the hectare study plot, 54 (27%) of the tree species and seven (50%) of the liana/strangler species were said to bear edible fruits. *Bertholletia excelsa* (the Brazil nut), for example, although unrepresented in the plot, is a common tree in the region and its 'nuts' are highly regarded by the Waimiri Atroari. The seeds are notably rich in oils (70-72% in the kernel) and in proteins (25%) (Pesce, 1985), as are those of the 'sapucaia' *Lecythis pisonis* (Lecythidaceae) whose seeds are also edible. Some of these edible species, particularly those with small sweet fruits, were said to be especially appreciated by the women.

One of the most diverse fruit tree families present in the region is the Sapotaceae, represented by 23 species within the hectare plot of which 14 were said to bear edible fruits. The majority of these are *Pouteria* species, of which *P. guianensis* and *P. cuspidata* seem to be particularly highly regarded. Although their use was not observed due to the season in which the study was conducted, the fruits of *Chrysophyllum prieurii* also appeared to be greatly appreciated, as were those of *Manilkara bidentata*, *Micropholis guyanensis* and *Ecclinusa ramiflora* etc. Another family of importance in this respect is the Moraceae, represented by several edible species of *Pourouma* (whose fruits taste rather like grapes), *Naucleopsis*, *Pseudolmedia* etc. *Caryocar villosum* (Caryocaraceae), like *Bertholletia excelsa*, is valued to such an extent that these trees are often left standing in swidden clearings on account of their fruits. The fruits, as well as providing a nutritious mesocarp (25.6% fat *fide* Aguiar *et al.*, 1980), contain seeds which can be roasted and eaten in the same way as are the seeds of *Osteophloem platyspermum* (Myristicaceae). Furthermore the bark may be used as a source of salt, which although unnecessary today as a result of the supplies brought in by Programa Waimiri Atroari, was of importance in the past. This was achieved by drying slabs of the bark thoroughly over a fire, burning them, and extracting the

* At least two other large-seeded legume species, *Mora excelsa* Benth. and *Voucapoua americana* Aublet, have been recorded as famine foods among the Indians of Guyana (Roth, 1924).

salt from the ashes with water. The fruits of the *Jessenia bataua* and *Mauritia flexuosa* palms would be used for the same purpose.

Fruits of the Palmae also furnish an important supply of nutrients, not only to the Waimiri Atroari but to Amazonians in general. The oily mesocarp of the fruits of *Jessenia bataua*, for example, a common palm of the damper regions of the forest, provides proteins which (in terms of amino acid content) are said to be comparable with those of high quality animal flesh, and 40% higher in 'biological value' (% retention of nitrogen) than soya proteins (Balick & Gershoff, 1981). The fruits of the equally abundant *Mauritia flexuosa*, which contain high levels of vitamin A (Aguiar *et al.*, 1980) and are available throughout most of the year, are much used in the preparation of nutritious beverages. Other common palm species of nutritional importance to these people include *Oenocarpus bacaba* (of which there were 24 individuals in the surveyed hectare), *Astrocaryum aculeatum* and *Euterpe precatoria*.

Many of the other plant families encountered during this study were represented by at least one wild species whose fruits are apparently edible. These included the Myrtaceae (e.g. *Myrcia guianensis*), Hippocrateaceae (e.g. *Salacia* spp.), Combretaceae (*Buchenavia grandis*), Humiriaceae (*Duckesia verrucosa*), Chrysobalanaceae (e.g. *Parinari montana*), Leguminosae (e.g. *Inga capitata*), Quiinaceae (*Lacunaria jenmanii*), Rubiaceae (*Duroia saccifera*), Apocynaceae (e.g. *Lacmellea ramosissima*), Annonaceae (*Ephedranthus amazonicus* (?)), Burseraceae (e.g. *Tetragastris panamensis*), Loganiaceae (*Strychnos jobertiana*), Melastomataceae (*Miconia serialis*), Menispermaceae (*Abuta grandifolia*), Passifloraceae (*Passiflora nitida*), Polygalaceae (*Moutabea* sp.) and Sterculiaceae (e.g. *Theobroma subincanum*).

It seems that the Waimiri Atroari do not eat leaves in significant quantities, and this may well be the case for Amazonian peoples in general. When offered a piece of white cabbage brought to Maré village by SRP, one of the Indians replied 'I don't eat leaves; I'm not a howler monkey!'. It is interesting to note, therefore, that the leaves of several crop species native to South America, including *Ipomoea batatas* (sweet potato) and *Anacardium occidentale* (cashew), are consumed in other parts of the world to which they are not native (see Table 3).

Although there appear to be many species of potentially edible fungi in the Amazonian forests, few tribes have been shown to exploit them to any significant extent (Fidalgo, 1965). Some notable exceptions, however, are the Yanomamo (Yanomami) Indians of northern Brazil, with whom Prance (1984) collected 21 species of edible fungi from just one village, all of which were known by different names. He noted that most of the fungi were collected from abandoned or 'fallow' swidden sites, and that as a result it was the women who tended to gather them and who possessed the most detailed knowledge of their properties. This may to some extent explain the paucity of mycological information offered by the Waimiri Atroari informants (exclusively male) during the present study, who claimed that there was just one species which they occasionally ate (not encountered), and that all other fungi (*panakry*) serve as nothing more than food for tortoises (note that Moskovits (1985), in her study of the ecology of the tortoises of northern Brazil, found fungi to be their preferred food item). Alternatively (and perhaps more probably, given that Baines (pers. comm.) was told that they eat none at all), it may be that fungi form no significant part of the Waimiri Atroari diet. The Maiongong Indians of northern Roraima, another Carib group, have also been reported to eat only one species of fungus, which they regard as an inferior famine food only to be used *in extremis* (Fidalgo & Prance, 1976).

Plants are used as sources of potable water in many parts of the world. Even in such a comparatively wet environment as tropical forest, these can be very useful on hunting trips or in temporary camps situated away from streams during the dry season. In the Neotropics, where some of the sources commonly employed elsewhere (e.g. bamboo and rattan) are generally sparse or absent, the most commonly used source is the woody stem of certain species of liana. The most productive stems are those which hang in loops to the ground, from the lowest point of which sections are cut and drained. Several plant families (including Vitaceae, Hippocrateaceae and Dilleniaceae) contain species which may be used for this purpose, and the Waimiri Atroari were found to use at least two: *Pinzona coriacea* (Dilleniaceae) and *Doliocarpus brevipedicellatus* (Dilleniaceae).

The Waimiri Atroari possess a remarkable degree of knowledge of the interrelationships between the fauna and flora of their forests. Much of this information, for obvious reasons, concerns the behaviour of the animals which they hunt for food. Of the 214 tree and liana species in the study plot, 141 (66 %) were recognized as food sources for game. The agouti, for example, is known to eat the seeds of *Clathrotropis macrocarpa* (in spite of the alkaloids which they contain), tapirs and tortoises eat the fruits of *Croton lanjouwensis*, and macaws, hunted for their feathers, eat the seeds of *Marmaroxylon racemosum* and sleep in its branches. Whether the fact that a tree provides food for game that are eaten by the Waimiri Atroari justifies its classification as a 'food tree', or as a 'useful tree', is debatable — and debated. It is not particularly important, however, provided that one recognizes that without these trees, without their pollinators, and thus without most of the rest of the food chain, life in the forest would not be possible for these Indians.

In certain cases trees will be actively 'used' as locators/attractants of game. This requires a combined knowledge of tree phenology and animal diet, and is particularly useful for nocturnal animals or those otherwise hard to locate. *Mixi* palms (*Mauritia flexuosa*), for example, are known to attract tapirs (amongst other game animals) when they drop their fruits, and platforms may even be built amongst them on which hunters will await their quarry by night. So although *mixi* may be felled for their fruits or their leaves, those standing close to a village will generally be spared for the attraction of game. Other species appear to act as phenological indicators. It was said, for example, that when the *dia* (*Osteophloem platyspermum*, Myristicaceae) bears fruits, then it is time to hunt the spider monkeys, who prior to this event are thin and unworthy of killing.

The only individual of *Astrocaryum aculeatum* (Palmae) encountered during the survey had been felled by the Indians for its fruits. Similarly, during the authors' visit to the village of Mynawa which coincided with the preparations for a large *marba* festival, a number of *Mauritia flexuosa* palms were felled alongside the BR174 highway for the preparation of beverages from their fruits. During the same visit a large specimen of *Parahancornia amapa* (Apocynaceae), heavily fruiting, was cut down in the vicinity of the village. Most of this felling was accomplished with chain-saws, which have recently been made available to the Indians by Programa Waimiri Atroari.

It appears that the felling of fruit trees is not an uncommon phenomenon amongst the Waimiri Atroari, although this may not have been the case in the past. Prior to the massive depletion of their population in the 1970s there would have been considerably greater pressure upon their resources than there is today, and the powers of renewal of these resources would have been of substantial importance to the Indians. In the vicinity of Iquitos (Peru), for example, the felling of *Mauritia* and *Jessenia* palms for their fruits has, within a comparatively short space of time, destroyed what was once a thriving market for these species (Vasquez & Gentry, 1989).

In the forests around the village of Maré were found the fallen trunks of *Hymenaea* sp. (Leguminosae), *Theobroma subincanum* (Sterculiaceae), *Oenocarpus bacaba* (Palmae), *Manilkara* sp. (Sapotaceae) and even the Brazil nut *Bertholletia excelsa* (Lecythidaceae). The felling of the latter species was particularly surprising, given that it tends to be left standing in swidden clearings (see Plate 3e), and that the fruits are generally not considered ripe until they fall. One (rather cynical) explanation which could be postulated for this apparent change in behaviour is that until the advent of the chain-saw, the felling of a Brazil-nut tree involved more work than it was worth. However, bearing in mind the recent history of the tribe and the policies that have been executed within their reserve, such behavioural changes may well be reflections of the rapidity of the technological advancement which these Indians have undergone, and of the consequential disorientation of their attitudes towards their environment.

Plants Used In Technology

Hunting

The principal weapon of the Waimiri Atroari is the longbow, made from the heartwood of *piria* (*Brosimum guianense*, Moraceae). The bow (*warpa*) may be as long as 2.2m, and is rectangular in section (see Plate 5a). *Piria*, also used for the same purpose by the True Caribs of Guiana (Im Thurn, 1883), is a highly prized commodity, and a broken bow will be kept and the pieces re-used for arrow-points. In the past it has been an important item of trade between Guianan and Brazilian Indians (Roth, 1924). Curiously, the Tembé of Pará and the Chácobo of Bolivia are reported to use this species for fuel (Balée, 1987; Boom, 1987), and many eastern Amazonian peoples use *B. rubescens* for their weapons instead (Balée, pers. comm.).

The bow is strung with a stout 3-ply cord made from the leaf fibres of a cultivated species of bromeliad. The leaves are cut at the base and split, and are pulled through a string noose in such a way as to macerate them, leaving a bunch of silky fibres. These are washed, dried, and rolled on the thigh (see Plate 5e) into the strong *karwa* string. An identical preparation method has been reported for the island Caribs of Dominica (Hodge & Taylor, 1957). The arrows (*pyrwa*) are made from the long straight inflorescences of *Gynerium sagittatum* (*mapyry*), a giant grass grown in large quantities in the swidden clearings to maintain a regular supply. The shafts are straightened by heating the irregularities over a low fire and gently flexing, and are then dried in the sun or over a cooking hearth. At the proximal end a small wooden plug (*pyrwa iasety*), with a nock to take the bow-string, is pushed into the pith. At the distal end of the arrow a longer 'stele' or fore-shaft of close-grained hardwood (*wiwe*) is added, its long tapering point twisted into the arrow-grass shaft whose end has temporarily been bound to prevent it from splitting.

These hardwood sections, which lend weight to the arrow and prevent it from breaking off once buried in the quarry, may be made from a number of tree species known collectively as *pyrwa iepry*. The 10 species within the study plot indicated as suitable for this purpose included close-grained species of Chrysobalanaceae, Myrtaceae and Leguminosae. According to Barbosa Rodrigues (1885), these fore-shafts were decorated by the Crichaná with the sap of a melastomataceous tree. This was not observed during the present study, but the red sap of various species of Melastomataceae including *Myriaspora decipiens* Naud. and *Miconia chrysophylla* (Rich.) Urb. is still used to decorate arrow tips by the Wayãpi Indians of French Guiana (Grenand, 1980). Simple or carved wooden points (generally of *piria*) may be used on these arrows (see Plate 5b and Fig. 5), but most will be fitted with a steel point (*pyrwa pytry*) and barb (*pyrwa iamera*), varying in shape and size to suit the intended quarry. According to Barbosa Rodrigues (1885), prior to the advent of steel the Crichaná (see p. 12) made these arrow-heads from the bones of their enemies.

The junctions between wood and shaft are secured using a *damymy* —a cord of *karwa* on two toggles — which is looped around the joint and pulled tight to compress the shaft around the insertion. They are then bound with *karwa* fibres and *manji*. *Manji* resin, although primarily used as an adhesive for arrow-

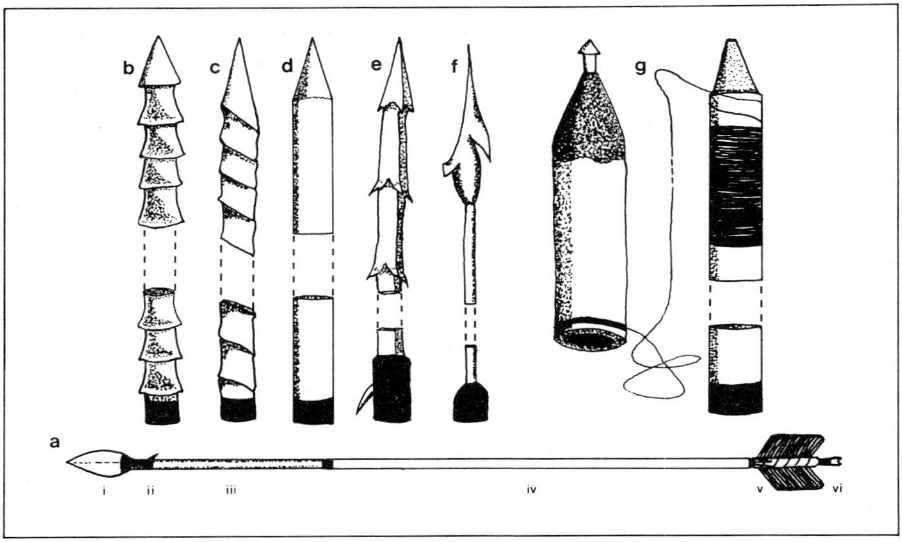

Figure 5. Contemporary Waimiri Atroari arrow construction: a) arrow, i) arrow-head (*pyrwa pytry*) of the *maia tadeha* type used for monkeys and smaller game, ii) steel barb (*maia iamera*), iii) fore-shaft or stele (*wiwe*), iv) shaft (*mapyry*), v) flights (*pyrwa iperiri*) (see also Plate 4c), vi) nock (*pyrwa iasety*); b–d) carved wooden arrow-heads made from *Brosimum guianense* wood and used mainly for birds and (b & c) for fishes, 30–60 cm long; e–f) barbed steel arrow-heads (*xere*) for fishing, 12–15 cm long; g) detaching arrow-head (*ta'kwa*) used for turtles — the head (left), once embedded in the carapace, remains attached to the floating shaft (right) by a thin cord, allowing the submerged animal to be hauled in.

making, may serve a variety of purposes. *Manji* is the name given to *Symphonia globulifera* (Guttiferae), a common forest tree whose glutinous yellow latex is bled from the bark. This is mixed with the boiled latex of *Couma macrocarpa* (Apocynaceae) or *Brosimum parinarioides* (Moraceae), and the resultant sticky mass moulded by hand (under-water) into solid cylindrical tablets. *Karwa* fibres are pulled through the *manji* block, much as cotton is pulled through beeswax elsewhere, lending strength and adhesion. The treated fibres are wound tightly around the joints, and the whole binding is coated with more *manji* which has been softened over a flame. *Manji*-coated *karwa* is also used to fletch the arrows (fix the feather flights to the shaft, see Plate 4c), and to attach the arrow-head*. The use of *Symphonia globulifera* for arrow-making, boat caulking etc. is widespread in Amazonia; the Indians of Guyana mix it with beeswax (Im Thurn, 1883).

Unlike many Amazonian tribes, the Waimiri Atroari do not use arrow poisons, though it was claimed by one informant that they had done so in the past but that the practice had since been rejected as dangerous. Whether or not this is the case, some of the requisite species are certainly available within the forest. Both *Virola theiodora* (Myristicaceae) and *Naucleopsis mello-barretoi* (Moraceae), used by the Waiká (Yanomami) and the Makú Indians respectively (Prance 1972), were present in the study plot. So also was a species of *Strychnos* used by the Kuripako Indians of Colombia (Altschul & Lipp, 1982). According to one informant the Waimiri Atroari had previously employed the copious red sap of the young *Machaerium quinata* vine (Leguminosae) for this purpose (boiled to a sticky consistency), but this was not corroborated by other informants and at present there is no pharmacological information to support or question the assertion.

However, poisons are still occasionally used by the Indians for catching fish, although this was acknowledged by Arie José as an indiscriminate process with harmful effects upon the environment. At Maré there was generally no shortage of fish to be speared or shot in the flooded igapó forests (at least during the high water period when most of this study was carried out), but at certain times of the year poisoning of streams or pools is resorted to. To what extent the use of fish poisons is traditional to the Waimiri Atroari is extremely difficult to ascertain; it seems that the use of at least one of the plant species employed has been learnt from the Waiwai Indians, with whom the Waimiri Atroari have been in contact since the 1960's or earlier. The use of such poisons by other Amazonian tribes is widespread and well-documented (e.g. Acevedo-Rodrigues, 1990; Killip & Smith, 1935). Most are extracted from the families Leguminosae, Sapindaceae and Euphorbiaceae (Acevedo-Rodrigues, 1990). Pieces of vine stem, tree bark, fruits or foliage are generally beaten with sticks and agitated in shallow water, whereupon the fish flounder at the surface and can be netted or shot.

Several plant species were said by the Waimiri Atroari to be toxic to fish, although their use was not observed by the authors. These include the two vines *Derris amazonica* (Leguminosae) and *Humirianthera rupestris* (Icacinaceae), and the four trees *Lecythis pisonis* (Lecythidaceae), *L. retusa*, *ma'xi* (which appears to be the most highly regarded poison) and *kaxiri*. The last two were not collected, but *ma'xi* was said to be a very large forest tree (probably Leguminosae), and

* Waimiri Atroari fletching follows Heath & Chiara's (1977) 'parallelogram with continuous radial open binding' pattern.

kaxiri, recorded elsewhere as synonymous with the vernacular 'jutaí pororoca' (Azevedo Lima, pers. comm.), is therefore possibly *Dialium guianense* (Aublet) Sandw. (Leguminosae). However, there appear to be no previous records of the use of this species for poisoning fishes, and this may warrant further investigation.

Two other species of vines, both present in the hectare study plot, were said by informants from the neighbouring village of Cacau to be used as fish poisons. These were *Maripa* sp. (Convolvulaceae) and *Strychnos jobertiana* (Loganiaceae). Although *S. jobertiana* does not appear to have been reported previously as a fish poison, a closely related species with a similar content of tertiary alkaloids (*S. rondeletoides* Spruce) has been attributed this use by the Paumaris Indians of Brazil (Altschul & Lipp, 1982). The use of *Derris amazonica* is likewise recorded for the Chácobo of Bolivia, although they regard this species as a comparatively weak poison (Boom, 1987). Its active constituent is rotenone, an isoflavonoid which blocks mitochondrial respiration and which has been found in the roots at a concentration of 0.3% (Krukoff & Smith, 1937). Rotenone is used for fish poisoning by scientific institutions, and is also marketed commercially as an insecticide.

Although tests on *Humirianthera ampla* (Miers) Baehni have shown it to be highly toxic to cattle, similar tests on *H. rupestris* have given negative results (Tokarnia *et al.*, 1979). The toxic constituent of *H. ampla* has not been isolated, but analysis of the tuber of *H. rupestris* has revealed the presence of degraded diterpenoids (Zoghbi *et al.*, 1981), and it is not inconceivable that these could be the agents responsible for the toxicity of this species. Several species of *Barringtonia* (Lecythidaceae) are used as fish poisons in the Far East (Howes, 1930), and one of these (*B. asiatica* (L.) Kurz) is actively cultivated for this purpose on the island of Martinique (Moretti & Grenand, 1982). There are, however, very few accounts of the application of Lecythidaceae for fish-poisoning in the Neotropics. *Gustavia augusta* L., in the seeds and bark of which saponins have been found (Grenand *et al.*, 1987), has apparently been used in Guyana in the past (Ernst, 1888). Saponins, which bring about irreversible changes in the permeability of fishes' gill cell membranes resulting in lethal losses of K^+ ions (Bissett, 1989), are the active constituents of many of the plant-derived fish poisons used in South America. They are present in high concentrations (15%) in the seeds of *Barringtonia asiatica* (Moretti & Grenand, 1982) and have also been identified in the bark of *Lecythis pisonis* (Elisabetsky, 1987), and it is therefore very likely that these are the compounds responsible for the toxicity of the *Lecythis* species used by the Waimiri Atroari.

A less effective poison, but one which is regularly available within the village, is 'tucupi' or manioc-root extract. Certain 'bitter' varieties of manioc (*Manihot esculenta*, Euphorbiaceae) contain cyanogenic glycosides which must be squeezed from the grated roots* prior to eating (the mean concentration of HCN in the manioc cultivars used by the Tukano Indians was found to be 1093ppm DM (Dufour, 1988)). This extract may be collected in large pans, and when a considerable quantity has accumulated, it is tipped directly into the

* The varying toxicity of manioc cultivars is discussed in detail by Nye (1991), who concludes that toxicity may not be the chief reason behind the complex processing systems employed in Amazonia.

stream. This method of fish-poisoning has also been recorded for the Witoto Indians of Colombia (Schultes & Raffauf, 1990).

Fish are also taken by the Waimiri Atroari in simple traps (*maxa*, Fig. 6), which are deployed in shallow streams. Traps of a very similar (open-ended) design are used by the Indians on the Uaupés River (Roth, 1924). They are made from split lengths of the *irwa* palm stem (*Iriartella setigera*), bound with the aerial roots of *ahnjaty* (*Heteropsis* sp.). The traps are secured with two sticks, with their openings facing the current, and the rest of the stream is blocked with a fence of palm fronds. These *maxa*, apparently little used nowadays, are said to be effective for catching 'traira' *Hoplias malabaricus* (Erythrinidae) and various cichlid fish species. Fish and caiman spears (the latter with detachable heads) may be made from *Duguetia flagellaris*, a straight-stemmed understorey treelet. For turtle fishing a heavier spear is required, normally made from *piria* wood and fitted with a short steel point.

Although many Amazonian tribes traditionally employ wooden war-clubs or 'bordunas', the Waimiri Atroari do not. It seems that the only clubs which they use are those for killing collared peccaries, cut from the heavy wood of various annonaceous trees. This process was not observed, but it was said that the peccaries, once they have been run to earth by dogs, are ringed with a fence of wooden stakes. One by one they are then chased from their burrows, and

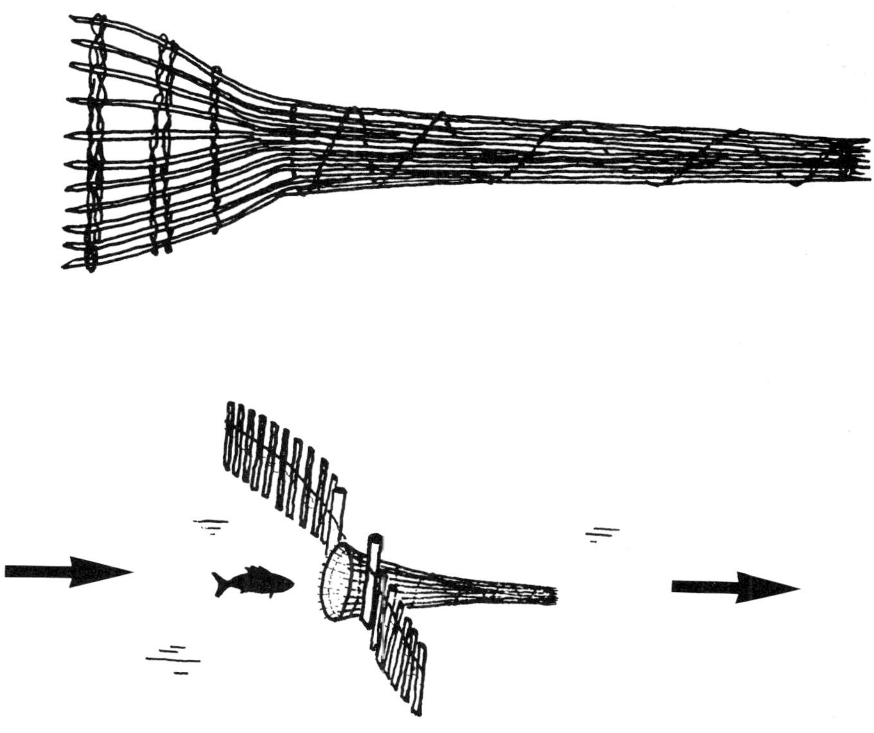

Figure 6. Waimiri Atroari fish-trap (*maxa*), and its method of deployment in a shallow stream (see above).

clubbed to death within the enclosure. If more are trapped than are needed, then they may be left in their prison until the occasion demands, and then evicted and slaughtered.

Craft

The Waimiri Atroari, like most Amazonian Indian tribes, are highly skilled in basketry. The principal material used is *baky* (*Ischnosiphon arouma*), although certain other species of Marantaceae will serve. The green stems are split into strips, the pith is scraped off, and each strip is then re-split into two layers. The rough inner layer is used for rough basketry or small animal cages (*xi'*), woven in a simple open hexagonal mesh (Plate 5b). The smooth outer layer is dried and kept for finer 'twilled' work by the men, such as panniers (*wiepi*), boxes (*pakra*, see Plate 4d) for storing arrow-making materials, sieves (*mare*, see Plate 5c) for making fruit drinks etc. and tubular manioc-squeezers (*matepi*). Complex designs representing aspects of the animal, plant, human and spiritual world, are woven into the *pakra* and *wiepi* using blackened strips of *baky*. To prepare the dye, resin (*wara'*) from the bark of certain *Protium* species (Burseraceae) is burned beneath a pot, on which a fine soot accumulates. Sections of bark are then cut from the trunk of a large forest *Inga* species, and their inner surfaces scraped of the soft cambium and sticky pink sap. This is mixed with the soot and daubed onto the inner (split) surface of the *baky*, where it becomes permanently fixed. Similar use of *Inga* as a fixative for lamp-black (in this case *I. bourgoni* DC.) has been recorded from French Guiana, where the mixture is used to dye gourds (Heim, 1903).

The *wiepi* (panniers) are mainly used by the women, for the collection of firewood and manioc etc. When fruits or small game are brought from the forest, a rough and disposable carrier is usually woven from two leaves of the *patwa* palm *Jessenia bataua* or another suitable species (these baskets are also used for soaking manioc roots in the river prior to grating). A carrying strap or tump-line, worn over the forehead, is stripped from the trunk of a tree. Many forest trees, particularly in the families Annonaceae and Lecythidaceae, possess fibrous inner barks (*amoty*) suitable for this purpose. In the hectare study plot 14 species were accredited with useful fibrous bark, including *Eschweilera coriacea* (of which there were 53 individuals present), *Rollinia insignis* (Annonaceae), *Apeiba echinata* (Tiliaceae), *Macrolobium unijugum* (Leguminosae), *Sterculia pruriens* (Sterculiaceae) and *Bombax* sp. (Bombacaceae). Hunters use these to tie up their quarry and carry it home, and thicker loops of fibres (generally from the Brazil-nut tree *Bertholletia excelsa*), are sewn into carrying straps for babies.

For the making of hammocks (*samka*), fibres from the leaves of the *mixi* palm *Mauritia flexuosa* are used (although it is not unusual today to find that over half of the hammocks in a maloca are of the woven cotton type, brought in from the city). The epidermis is stripped from the surfaces of young unopened pinnae, dried (see Plate 5f), and rolled into a string on the thighs. The hammock itself (see Plate 2) is then woven on a simple 2-pole frame. Identical methods of hammock-making were described by Im Thurn (1883) for the Arawak Indians of Guyana, but many Amazonian tribes employ the durable fibres of *Astrocaryum* palms for this purpose instead. The Waimiri Atroari also collect the unopened pinnae of the *sykma'* and *kwirba* palms (*Astrocaryum* spp.), but use them to weave their fans for cooking fires and to make mats.

Plate 2. Waimiri Atroari girl at Mynawa village in 1989, lying in a hammock (*samka*) made from the fibres of the 'buriti' palm *Mauritia flexuosa* (*mixi*).

Pottery is an art which seems virtually to have been lost by the Waimiri Atroari since the advent of aluminium pots and pans. Although there has since been something of a resurgence (for trade rather than use), at the time of study it was apparently being practised by one woman only, who lived at the Xara village on the Alalaú River (Fig. 1). According to informants at Maré, the best clay was obtained from the bed of the river, for which (at that time of year) one had to dive. This was mixed with the ashes of the bark of certain species of *Licania* (Chrysobalanaceae), which are rich in silica and help to strengthen the pot. Crystals of silica are present (to varying extents) within the trunks of a great variety of South American trees (Pernia, 1982), and their use for this purpose is widespread and well documented amongst Amazonian Indians (e.g. Roth, 1924; Prance, 1972). The two species indicated as suitable by the Waimiri Atroari were *L. longistyla* and *L. micrantha*, but some of the other species present in the hectare study plot have been attributed equivalent uses elsewhere in Amazonia. The pottery would generally be coated with a glaze of the 'copal' resin which exudes naturally from the trunk of *Hymenaea* spp. (Leguminosae), and would be decorated with vegetable dyes (Barbosa Rodrigues, 1885).

Adornment

The Waimiri Atroari use little in the way of personal adornment. Traditionally the men would wear a length of an araceous aerial root wrapped several times around the waist to support the penis (see Plate 1), and the women a small 'tanga' (*kwyie*) which hung in front of the pudenda, supported by a *karwa* (bromeliad fibre) string. Neither of these are still in use, having been substituted by modern clothing. The *kwyie* was made of split *paxiy'* seeds (*Socratea exorrhiza*) and another round black seed (unidentified — Palmae?), and sometimes the woody bases of the male flowers of a hemiepiphytic *Clusia* species (*marabia*), sewn together with *karwa* thread. Very similar items were described by Barbosa Rodrigues (1885) amongst the Crichaná:

'*As mulheres andam tangadas com* umaiṕos *de fructas de* puća (Cissus *sp.) tecidos com fios de fibras, tendo uma franja feita de albumen do Caraná e endocarp de Marajá, partidos no meio, enfeitada com ossos de ouvido de cotia e calices de uma flor que penso ser uma Myrtaceae*'.

Bracelets (*warma*) are made from the stems of *Ischnosiphon polyphyllus* (*warmaxi*), which are scraped, split into thin strips, and then skilfully plaited into large continuous loops (Plate 4b). These are wound several times around the wrist to form a broad band, whose original function was probably as a 'bracer' to protect against abrasion from the bow-string (a leaf of *karwa* (Bromeliaceae) may be tied around the wrist for the same effect). Barbosa Rodrigues (1885) noted similar wrist bands, called *larumaty,* on his visits to the Crichaná in the last century.

At the *marba* festivities the body may be painted with *manji* latex from the bark of the *Symphonia globulifera* tree (Guttiferae), which stains the skin dark brown and lasts for several days, or with the more common dyes (in Amazonia) prepared from *Genipa* sp. (Rubiaceae) and *Bixa orellana* (Bixaceae). Head-dresses, also worn at these ceremonies, consist of simple head-bands woven from *Ischnosiphon arouma,* interwoven with (horizontal) hawk or eagle breast feathers and with a few macaw or eagle flight feathers inserted vertically.

Transport

Traditionally the Waimiri Atroari built large wooden dug-out canoes, as do a great many Amazonian peoples. These boats, to which there are numerous references in the early documents concerning the tribe (e.g. Barbosa Rodrigues, 1885), have largely been superseded by the aluminium 'voadeira' canoes (with outboard motors) which are now provided by the Programa Waimiri Atroari. In fact the building of these traditional dug-outs has been actively discouraged in the past by FUNAI employees, on the grounds that they are 'no good' (Baines, 1988). For fishing with bow and arrow in the flooded igapó forest, however, aluminium boats are unsuitable on account of their size and lack of manoeuvrability. The ideal vehicle for this type of hunting is a short light dug-out, in which one can approach the feeding fish with scarcely a ripple or a noise. During the period of the present study there was a marked shortage of these boats at the village of Maré, although at that time fish provided the major source of protein in the Indians' diet. When asked why they did not build more canoes, however, they replied that only the old men still knew how to do so. This reply probably owes more to the past discouragement of the continuation of traditional practices than to a genuine lack of the necessary technology, but whatever the case the ironic result was that in 1989 a canoe-builder was employed by Programa Waimiri Atroari to work on the Indians' behalf.

The Waimiri Atroari did not light fires under their dug-outs to open them out (as is often practised elsewhere), as a result of which they were rather narrow and unstable (Carvalho, 1982). During this study 10 tree species were noted as suitable for canoe construction, including five species of Lauraceae. The Lauraceae, particularly *Mezilaurus* spp., are utilized throughout Amazonia for the construction of canoes and larger boats, on account of their strength and resistance to rot. According to Barbosa Rodrigues (1885), the Crichaná made their dug-outs ('curiara') from the trunk of *Anacardium brasiliense* Barb. Rodr. (Anacardiaceae). Interestingly, one of the species said to be useful for canoe making by the Waimiri Atroari was *Scleronema micranthum* (Bombacaceae), which is used by the Carib Indians of the upper Barama, Guiana, for bark-skin canoes (Roth, 1924). There are, however, no records of the use of this type of canoe by the Waimiri Atroari, and it may be that the information originates elsewhere (as appears also to be the case for the use of this species for bark walls — see p. 40).

Paddles were originally cut from the large thin buttress roots typical of many species of *Sloanea* (Elaeocarpaceae), but recently modern saws have eased the process of plank-making, and other woods such as *Mezilaurus* sp. and *Lueheopsis rosea* may be used instead (see Plate 6b).

Miscellaneous

Plants are used by many Amazonian tribes to adorn the body in order to bring luck in the hunt. The Tembé Indians of Brazil, for example, wear bracelets of the root bark of *Sagotia racemosa* Baill. (Euphorbiaceae) on their wrists for this purpose (Balée, 1987). In addition, plants are used to boost the efficacy of their hunting dogs, a practice that presumably developed since the conquest of the Americas when *Canis familiaris* was first introduced. The Indians of Guyana, for example, force their dogs to eat pieces of *Caladium* spp. (Araceae), and push their noses into holes containing burning tapir hooves and biting insects (Im Thurn, 1887). To this end the Waimiri Atroari may persuade their dogs to sniff the crushed roots of *Piper consanguineum* (Piperaceae), or the inner bark of *Duguetia cauliflora* (Annonaceae), and they cut off the dogs' ears and tails etc. if they exhibit cowardice in the face of a peccary (*'pour encourager les autres'*). The leaves of immature individuals of the *paxiy'* palm *Socratea exorrhiza*, when crushed in water, produce a gelatinous substance which serves as soap or shampoo.

Medicinal Plants

A great deal has been written about the medicinal knowledge of the Indians of South America, and of its potential for the world at large. This is the aspect of ethnobotany which attracts the most public interest, the most discussion, the most money, the most controversy and the most rhetoric. Without doubt many of the plant species used medicinally by these peoples are so employed because they contain pharmacologically active chemicals which effect the required cure. It should not be forgotten, however, that in most if not all such societies, there are also plants whose 'medicinal' properties are attributed to them through the type of morphological association known as 'doctrine of signatures'. To separate the two is not necessarily a simple matter, or as Davis & Yost (1983a) put it, 'In no area of ethnobotany is the challenge greater than in the search for new medicines, for in no area is there a greater mixture of fact and fiction'.

The degree to which Amazonian Indians have developed their medicinal plant knowledge varies greatly from tribe to tribe. The Waiká (Yanomami) Indians of northern Brazil, for example, rely heavily upon magic and shamanism and use little in the way of actual plant medicine (Prance, 1972). Of the plant species collected during this survey, 59 were attributed medicinal properties by the Waimiri Atroari including 15% of the tree species in the hectare study plot. Many of the species cultivated by the Waimiri Atroari (see Table 1) have consistently been attributed medicinal values by other Amazonian tribes (see Table 3), and although no such investigation was carried out during the present study, it is highly likely that some of these are likewise employed by the Waimiri Atroari.

A list of the medicinal plants, arranged according to the ailments for which they are used, is presented in Appendix IV. It is very often the bark of the plant which is used, and many of these barks are distinctly bitter to taste. Some of the ailments for which they are employed would not have been present in the Waimiri Atroari population prior to their contact with Europeans, and yet the unstable nature of that contact may have precluded the transfer of medicinal technology from the colonizing population until comparatively recently. It is a possibility, therefore, that some of the cures for those alien diseases have been independently developed by the Waimiri Atroari since their introduction.

At least 45% of the plants categorised as medicinal in this survey (and identified to the level of species) have been attributed medicinal properties by other tribes or peoples, and of these several were used for the same or similar maladies. The use of *Piper consanguineum* (Piperaceae) to treat wounds, for example, has also been recorded for the Chácobo of Bolivia (Boom, 1987), and the use of *Bauhinia guianensis* for excessive menstrual bleeding is also practised by the Kayapó of Brazil (Elisabetsky & Posey, 1989), although in a different manner. Of the medicinal plants listed here which have undergone chemical analysis, a number have been shown to contain compounds (often alkaloids) whose pharmacological properties or potential properties mirror those required by the applications to which the Waimiri Atroari put them.

It is interesting, although not altogether surprising, to note that in many cases a plant used medicinally by one tribe may well be employed as a poison by another (this is discussed in greater detail by Bissett (1989)). The bark exudates of *Naucleopsis mello-barretoi* (Moraceae) and of *Virola theiodora* (Myristicaceae), for instance, are used directly as arrow poisons by the Makú and Waiká Indians (Prance, 1972), but were said to be applied to wounds by the Waimiri Atroari to prevent and cure infection. So also was the latex of *Helicostylis tomentosa* (Moraceae), although in Colombia this is regarded as a toxic substance (Schultes, 1983). In a similar fashion the Waorani Indians of Ecuador use *Curarea tecunarum* Barneby & Krukoff (Menispermaceae)* as the principal ingredient of their potent arrow poison, but apply this poison directly to bacterial or fungal infections of the skin with demonstrable effect (Davis & Yost, 1983b). There appears to exist a similar correlation between plants used as hallucinogenic drugs and those used as vermifuges, indeed it has been suggested by Rodriguez & Cavin (1982) that the Amazonian hallucinogenic plants containing indole or isoquinoline alkaloids (principally the Myristicaceae) may originally have been selected by the Indians for their anti-parasitic properties.

One of the major concerns voiced by ethnobotanists (and indeed one of the main justifications for their existence), is the impending loss of the knowledge which they are recording. Unfortunately the Waimiri Atroari are no exceptions to this phenomenon, and it seems that one of the aspects of their cultural heritage most vulnerable to erosion is their knowledge of plant medicines. In the last 30 years the Waimiri Atroari population has been very seriously depleted by introduced diseases, notably measles. Since 1981–2, when the population was at its lowest ebb, medical support has done much to prevent further epidemics and the population has expanded. It is not surprising, therefore, that the Indians are now very enthusiastic about the powers of modern medicines, to the extent of neglecting some of their own traditional cures for their native ailments. This enthusiasm for modern drugs had until 1990 been in no way discouraged.

The knowledge of these cures will always have rested predominantly with the elders of the villages, as was clearly the case in Maré. The two men who were reputed to know the most about the subject were Warpeki Comprido and Kwaxi Manelão. Warpeki was the Waimiri Atroari equivalent of 'shaman' (i.e. spiritual leader) in the village, and it was he that would converse with the spirits at night, and who was called to a neighbouring village during the present study to assist a woman who was having a difficult childbirth. Kwaxi was the oldest man in the village, and provided much of the information presented in this paper.

It appears, however, that the distraction of an alternative to plant medicine, and the past influence of FUNAI officials within the area, have dulled the

* This species is also employed by the Dení Indians of Brazil as a contraceptive drug (Prance, 1972).

Plate 3. a. *Terra firme* forest in the hectare study plot. Note the importance of palms in the lower understorey. 79% of the 200 tree species in this plot were attributed useful properties by the Waimiri Atroari (see p. 117).

b. The Rio Alalaú — principal river of the current Waimiri Atroari territory and extremely important as one of their main sources of protein (fish, caiman, turtles etc.), and as a means of communication. Note the blackness of the water, resulting from the virtual absence of silty sediments and the high concentrations of tannins leached from rotting vegetation.

c. Maré — the village where the majority of the present study was carried out. The round house ('maloca') on the right is in the traditional style of the Waimiri Atroari, although smaller than some of the houses that were observed in the 1970's (see p. 38). Details of its construction are given on p. 39, and in Fig. 7. The buildings on the left include the house of Arie José (FUNAI-appointed 'captain' of the village), a new school, and a house for the processing of manioc. Banana plants (part of an old plantation that has been converted to pasture) are growing in the foreground.

d. View from the village of Curiaú on the Rio Curiuaú, showing seasonally flooded igapó forest in the background. Note the elevated situation of the village (see p. 11), and the planted fruit trees surrounding it (see p. 22). The palm on the right is the peach-palm *Bactris gasipaes*. Leaning against the palm are poles cut from the forest understorey.

e. Cleared land at the village of Maré. The large tree in the centre is a Brazil-nut tree *Bertholletia excelsa*, which has been left standing for its fruits, and the tree on the left (partly obscured by thatch) is the butter-nut tree *Caryocar villosum*. This clearing is much larger than would traditionally have been made by these Indians (see p. 9), and much of it was being converted to pasture at the time of study. The photograph is taken from the house of Arie José. The floor is made from the split trunks of the 'paxiuba' palm *Socratea exorrhiza*, which is apparently not a traditional Waimiri Atroari building material (see p. 40), and the uprights are of the olacaceous *Minquartia guianensis* (note the characteristically lobed trunk).

Plate 3. a b

c

d e

Plate 4. a. Three of the principal informants who contributed to the study at the village of Maré (see p. 15), discussing the properties of a pressed and dried specimen of *Heisteria* (Olacaceae). The man on the left is Warpeki Comprido, in the middle is Pawyna Elias and on the right Arie José. Note the wrist-watch of the latter, given by FUNAI to reinforce the status of village 'captain' which they have assigned to him. The shirt worn by Warpeki Comprido was handed out as propaganda during a recent election in Manaus. The palisade walls of the house, made from the wood of *Guatteria olivacea* (Annonaceae) and bound with the aerial roots of *Heteropsis* sp. (Araceae), were traditionally built to repel spear attacks (see p. 39).

b. Waimiri Atroari women of the village of Mynawa making *warma* wrist-bands from the stem of *Ischnosiphon polyphyllus* (Marantaceae). Traditionally these would probably have been used to protect the wrist from the bowstring (see p. 34). The glass beads worn on the woman's wrist have been obtained through exchange with the Waiwai Indians (see p. 11). The girl in the background is lying in a traditional hammock woven from the leaf-fibres of the *Mauritia flexuosa* palm. Note the plyboard walls of the house behind (see p. 40 for explanation).

c. Detail of the fletching of a Waimiri Atroari arrow (see p. 29). The stem is from the cultivated grass *Gynerium sagittatum*, the fibres binding the feathers and the nock are taken from the leaves of a cultivated bromeliad, and the black adhesive *manji* 'resin' is prepared from the latex of the wild forest trees *Symphonia globulifera* (Guttiferae) and *Couma macrocarpa* (Apocynaceae) (see p. 29). The upper flight feather is from the blue-and-yellow macaw *Ara ararauna*, the lower from the black curassow *Crax alector*, and the decorative feathers on the left are from the tail of a toucan. The fish in the background are 'traira' *Hoplias malabaricus*, which have been poisoned with the juice extracted from manioc tubers and then shot (see p. 30).

d. Traditional *pakra* double-layered and lidded basket at Mynawa village, used by the men for storing feathers, *manji* resin and tools for arrow-making. These skillfully decorated baskets are woven by men from the split stems of *Ischnosiphon arouma* (Marantaceae), some of which have been painted with a mixture of fine soot and resin (see p. 32). It is common practice for such items to be suspended from beams within the maloca, or (in the case of bows and arrows) pushed between the beams and the thatch. Elevated storage racks made of split *Iriartella* palm stems (see right of picture) may also be used.

Plate 4. b d

a c

Plate 5. a. Indians from the village of Mynawa, fishing with bows and arrows in a swamp close to the BR174 highway (see p. 7). Note the great length of the hunting bow, which is made from the wood of the moraceous *Brosimum guianense* (see p. 27), and the thickness of the bromeliad-fibre bowstring. Arrows are carried in the same hand as the bow, and not in a quiver. The man in the background is carrying a steel knife in his shorts, a practice commonly adopted by the Waimiri Atroari men, who are highly skilled hunters.

b. Warpeki Comprido carving the point for an arrow from the wood of a species of Myrtaceae. Certain tree species, generally with very fine-grained wood, are regarded as particularly suitable for this purpose (see p. 28). The wood is being cut with a steel knife, but in the past a tooth of the white-lipped peccary would probably have been used for this purpose. Note the construction of the roof and the thatch (see also Fig. 7), and the typical open-weave basket on the left in which animals may be kept or transported (see p. 32).

c. Women of Maré village preparing a beverage from bananas. The pulp is being forced through a sieve woven from *Ischnosiphon* stems (see p. 32). The pans are made from aluminium, which are more practical than the traditional ceramic items (see p. 46) and are now universally used by these people. Note the calabash at the bottom left of the picture (made from the fruits of the bignoniaceous *Crescentia cujete*), the heavy wooden mortar in the foreground, and the fire in the background over which meat is being smoked.

d. Woman of Maré village baking *mei* cakes, prepared from the washed and grated tubers of manioc *Manihot esculenta* (see p. 22). These form the main staple of the Waimiri Atroari, and are generally eaten with pieces of smoked meat or fish. Originally these cakes would have been baked on clay, but the introduced steel pans (now used throughout the Amazon) are a considerable improvement on this technique. Certain tree species are specifically revered as sources of good fuel-wood for cooking manioc, which requires a hot flame (see p. 41). Note the use of palm leaves for wrapping the bundle on the ground.

e. Preparation of string from the leaf-fibres of a cultivated bromeliad species at Maré village (see p. 27). The fibres are rolled on the thigh into a tough cord, which is used for stringing hunting bows and for suspending hammocks etc. This task appears usually to be the preserve of women, but evidently men are also conversant with the technique.

f. Fibres stripped from the young (unopened) leaves of the 'buriti' palm *Mauritia flexuosa* (see p. 32). When they have finished drying in the sun, they will be rolled into a cord (using a technique similar to that shown in the adjacent photograph) from which traditional string hammocks will then be made (see Plate 2).

Plate 5. a b

c d

e f

Plate 6. a. The old village of Taquari close to the BR174 highway, abandoned when much of the surrounding land was inundated by the Balbina dam (see p. 8). Traditionally the abandonment of a village would usually be related to the presence of bad spirits (associated with a decline in crop production or hunting success), and the houses would be burned. This particular village is made up of a number of small circular and rectangular houses, probably as a result of the influence of FUNAI employees who were operating in the area at the time of its construction (see p. 40). Note the typical colonizing plant species (e.g. *Cecropia*, *Solanum*) which are moving into the old clearing.

b. Arie José sharpens the blade of a plane, which he is using to finish a paddle from *tawyba* wood (*Mezilaurus* sp.). The plank for the paddle has been cut from a tree trunk with a chain-saw, whereas in the past it would have been cut from a thin buttress root. The advent of modern tools and technology has done much to facilitate the lives of the Waimiri Atroari, but the speed of their arrival poses serious problems (see p. 47). Note the large Brazilian-style manioc press in the background, which is sometimes used instead of the traditional woven manioc squeezers.

c. Part of the Pitinga tin mine (subsidiary of the Paranapanema mining corporation), on land which was once part of the Waimiri Atroari territory (see p. 9). A stream has been dammed for the dredging and washing of tin ore (cassiterite) from the valley floor deposits. The whole of the current Waimiri Atroari reserve is underlain by valuable tin deposits, which pose a serious threat to the future of these people (see p. 120).

d. Pollution from the Pitinga tin mine (the brown water in the background), entering the Alalaú river from its tributary the Tiarajú. There are three villages of Waimiri Atroari Indians on the Alalaú, and it has been reported that the pollution has had a detrimental effect both on their health and on their fish stocks (see p. 9). According to the legal statute which created the reserve, the entire basin of the Alalaú is contained within the reserve and such pollution should be impossible.

e. Young Waimiri Atroari boy inspecting the cattle at the village of Maré. Note the bow and arrow which he is carrying; the children regularly use these around the villages and by the time they reach maturity (and are ready to hunt), they are already adept at shooting. The introduction of cattle (see p. 119), however, threatens to upset the Indians' traditional relationship with their environment in a manner which is likely to be detrimental to their welfare. Note the scars on the boy's back, which are relics of his recent *marba* 'initiation' ritual (see p. 43).

f. One of the cattle ranches that have been set up on the Indians' behalf close to the BR174 highway. The project was not tended by the Indians themselves, but by a 'gaucho' cowhand who has been brought in from elsewhere. In the background newly cut forest is being burned to create fresh pasture, and old pasture is being burned to reduce weeds and encourage sprouting of new grass.

Plate 6. a b

c d

e f

enthusiasm of the young to inherit this knowledge through the oral tradition by which it has been conveyed in the past. This was clearly demonstrated by a poignant and very sad incident which took place during the period of this study. A group of Indians from another village, invited to participate in a *marba* ceremony, brought chicken-pox with them which rapidly swept through the population of Maré village. Kwaxi Manelão caught the disease, developed pneumonia, and died. Shortly afterwards Are'bri Romeo, a young man with little apparent interest in the trappings of our 'civilization', voiced his concern thus: 'Manelão has died; now *ki'nja* (Indians) no longer know about medicine'.

Evidently such 'deposition' of traditional information with specific members of the community, which is common amongst societies with oral traditions, renders it extremely vulnerable to loss. Pierre Grenand, for example, visited an isolated group of Wayãpi Indians on the Kouc river (Brazil) in the 1970's; a group which had recently suffered a serious epidemic (of grippe) which had killed several members of the community. He found that as a result of this one brief (but disastrous) episode, these people had lost more of their traditional knowledge than similar groups of Wayãpi in French Guiana had lost over fifteen years of contact with the French and Créoles (Grenand, pers. comm.).

Schultes (1979b), reporting on the Alpha-Helix Amazonian Expedition, cites an example which clearly emphasises the rapidity with which detectable losses of information of the kind described above may occur. He was working with the Bora and Witoto Indians in Colombia, collecting the myristicaceous hallucinogens which for two generations they had not used. The younger members of these tribes enthusiastically demonstrated to the biologists their traditional methods of collection and preparation of these snuffs from the bark-resins, apparently having remembered all in detail. But when the samples were subsequently analysed, many were found to have been taken from the wrong species, lacking the alkaloidal components necessary for the snuff's efficacy.

Plants Used In Construction

Temporary shelters used by the Waimiri Atroari in the forest may consist of nothing more than a bunch of *Jessenia bataua* palm fronds lashed together and planted in the ground, or secured at the base of a tree, but long-term accommodation generally takes the form of large communal thatched wood-frame round-houses. Indeed it has been said that the Waimiri Atroari built the largest such houses in the whole of Amazonia — up to 50 metres across — and although those extant today are considerably smaller than that, they remain impressive structures. The Waimiri Atroari possess a comprehensive knowledge of the wood properties of the trees in their forest, much of which is presumably gained from first-hand experience during the manual clearing of land for cultivation. Certain species, however, are traditionally revered for their strength and rot-resistance, and have specifically been employed in maloca (house) construction in the past. Of the 200 tree species present in the hectare study plot, 32% were recognized as suitable for this purpose. For the majority of the components a variety of these species will serve, but for the stakes which form the outer wall of the house (*iara*, see Plate 4a), one in particular was said to be preferred. This, *Guatteria olivacea* (Annonaceae) or *amoty*, is regarded as especially resistant to lances, which was presumably an important consideration before the discontinuation of inter-village skirmishes or inter-tribal wars. If necessary this may be effectively substituted by *warxi* (Myristicaceae).

A brief survey was made of the species used in the construction of the Maré maloca (Fig. 7). The main uprights within the house (*wykapy* and *mita wykapry*) were almost entirely of *Minquartia guianensis* (Olacaceae), also known as *kamia wykapry* or 'white man's house-post'. This is a strong and rot-resistant wood that is so respected by the Tembé Indians of Pará (Brazil) that there is a taboo on its use for firewood, the breaking of which is said to result in numerous deaths in the village (Balée, 1987). The outer ring of uprights (*asa wykapry* or *masna wykapry*), structurally less important, were made from a variety of trees including *Pouteria* spp., *Licania* spp., *Eschweilera rhododendrifolia*, *E. coriacea*, *Theobroma subincanum*, Lauraceae (various), Meliaceae (various) etc. For the radiating rafters (*ie'* or *iuwe'*) long straight poles are required, which is a characteristic of many annonaceous trees. At Maré these rafters included *Duguetia* sp., *Xylopia* sp., *Guatteria* sp. and *Bocageopsis multiflora*, as well as certain Myristicaceae and Leguminosae (*Tachigali/Sclerolobium*). Tie beams (*patne'ne*) included *Duguetia* sp., *Eschweilera* sp., *Pouteria* sp. and *Quiina* sp., and the thatch supports (*iara bahnja*) included *Eschweilera* spp., *Duguetia flagellaris*, *Pourouma* spp. and *Cecropia* sp.

The components were bound with *ahnjaty*, the tough and flexible aerial roots of araceous climbers (*Heteropsis* spp.) which abound in the forest. These roots, known colloquially in the region as 'cipó titica', are widely used by Indians and settlers for binding and for basketry etc. At Maré, *ahnjaty* was also used to bind the split sections of *Iriartella setigera* which make up the maloca doors (*mydy pari*).

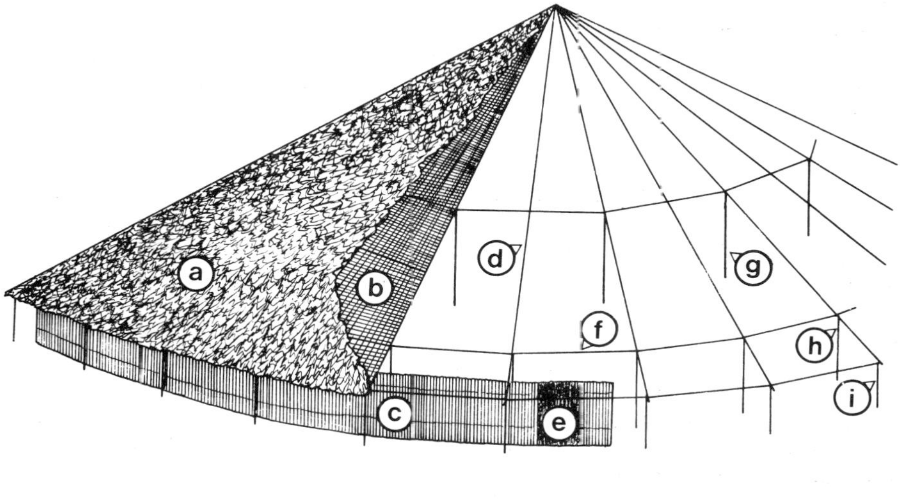

Figure 7. Schematic representation of the construction of the principal Waimiri Atroari house (*mydy*) at Maré village (see also Plate 3c): a) thatch; b) thatch support (*iara bahnja*); c) palisade (*iara*); d) rafter (*ie'*, *iuwe'*); e) door (*mydy pari*); f) tie beam (*patne'ne*); g–i) house-posts (*wykapy* — *mita wykapry*, *asa wykapry* & *masna wykapry*).

The choice of materials used in Waimiri Atroari maloca construction must depend upon both personal preference and local availability, and no two villages are totally alike. Maré, for example, is thatched with the leaves of the *wy'* palm *Manicaria* (?) *martiana*, which is said to grow in abundance further up the Rio Camanaú and was brought down by canoe. Many of the other malocas are thatched with *karbana* leaves (*Mauritia carana/M. aculeata*), collected from nearby swamps and 'caatinga' forests. Similarly, whereas the peripheral wall of the Maré maloca was built of *amoty* stakes, thatched and partially thatched walls were observed in other villages, and at Velho Taquari, a village abandoned by its inhabitants when much of the adjacent forest was flooded by the Balbina dam's waters, there were walls made from the bark of *Scleronema micranthum* (Bombacaceae). The bark had been stripped from the trees in its entirety, opened into broad panels, and then 'sewn' to the inner posts with *ahnjaty* (see above). This type of wall construction is apparently not traditional to the Waimiri Atroari but was introduced by FUNAI employees (Baines, pers. comm.). Identical use of the bark has been recorded among the Indians of Guyana (Roth, 1924; Fanshawe, 1950).

Curiously the village at Velho Taquari had not been built in the traditional Waimiri Atroari manner, but, as a result of outside influence, consisted of numerous small rectangular or round huts (Plate 6a). All of the Waimiri Atroari settlements have at some point passed through this small hut phase (Baines, 1988), and some of them have remained as such. When the Indians left Velho Taquari to build a new maloca, further from the road and from the flood waters, they were encouraged to revert to the traditional style. However the Eletronorte electricity company, as a gesture of 'compensation' to the Indians, modified the building by installing a cement floor and plywood walls (Plate 4b).

The effect of outside contact upon Waimiri Atroari choice of construction materials is hard to assess given the lack of previous information. The split trunk of the *Socratea exorrhiza* palm (see Plate 3e), for instance, is strong and flexible and is much used for walls and floors in Amazonian rustic architecture, but does not appear to be a component of traditional Waimiri Atroari buildings. Nevertheless, when a new head of the FUNAI post arrived at Maré in 1989 (bearing gifts such as plastic suitcases and jeans), and requested that the Indians dig him a new latrine pit and erect a hut over it, they made a long boat trip up the Rio Camanaú to an area of that species' abundance, and brought back the necessary wood to build the hut.

In December 1989 the Waimiri Atroari of Mynawa village were encouraged to build an experimental hen-house. At Mynawa a certain species of *Manilkara* (Sapotaceae) known as *parta*, which grows in abundance in the low-diversity 'caatinga'-type vegetation (low stature forest on sandy soil) some distance from the village, is particularly esteemed for its properties as a house-post. Although it seems likely that the obstacles of distance and transportation would traditionally have deterred them from utilizing this resource (they would have used a mixture of species from the adjacent forest), those problems were surmounted by the availability of a vehicle and a road (the BR 174). As a result, the hen-house was exclusively built from *parta*. The fact that so many species of trees were employed in construction of the house at Maré may be due to the high diversity of the surrounding forest (see Fig. 8), and the resultant low distribution densities of most of its component species. Those trees regarded as the most suitable will probably be sufficiently sparse in the region to make their exclusive use impractical. A group of Venezuelan Panare Indians recognized only 2.9% of the species in a studied hectare of forest as suitable for construction (Boom, 1990), and this may either be a reflection of the lower forest diversity (70 spp. in a hectare as opposed to 200), or of a particular local abundance of their chosen species.

Fuel Plants

The suitability of a tree species for fire-wood has not, in this study, been categorized as a useful property. Fuel is obviously of great importance to the Indians, but for forest-living peoples its superabundance prevents it from significantly influencing their lifestyles. Most tree species can be used for fuel, as was shown by Balée's studies with the Tembé and the Ka'apor. These revealed potential firewood utilization of 91% and 82% respectively for the species of trees in one-hectare study plots (Balée, 1986; 1987).

The Waimiri Atroari appear to take most of their firewood from their swidden clearings, its collection being the responsibility of the women. Many of the larger trunks will have survived the burning of the felled forest, and once dried in the sun provide an ideal source of fuel. Particular species are regarded as specifically suited to particular purposes, due to their differing burning properties. *Goupia glabra* (Celastraceae), for instance, was classified as *wyty iaty* (meat firewood). Meat and fish are cooked over a smoky fire which is maintained until all is eaten — perhaps several days — so a fast-burning wood or one which releases acrid or poisonous smoke will clearly not be suitable*. *Eschweilera rhododendrifolia* (Lecythidaceae) was classified as *mei iaty*, or 'manioc firewood', and other species noted as good fuel-woods included *Protium* spp., *Licania* spp., *Eschweilera coriacea*, *Guatteria* spp., *Jacaranda copaia*, *Apeiba echinata*, *Lueheopsis rosea*, *Theobroma subincanum*, *Scleronema micranthum*, *Clathrotropis macrocarpa* and *Rinorea racemosa*. This list is by no means exhaustive. *Pourouma ovata*, whose sawdust was said to cause skin irritation, is avoided as a firewood because the smoke is also reputedly an irritant. Conversely the wood of *wyrawyra* (*Ferdinandusa* sp., Rubiaceae), even when freshly cut, is reputed to burn so well ('like gasoline') that Indians travelling through the forest in wet weather will actively procure it. *Wyrawyra* is said to have been much used by the Waimiri Atroari ancestors.

The use of resins from the bark of burseraceous trees (chiefly *Protium* spp.) is widespread throughout Amazonia. These resins, often exuded from natural wounds in considerable quantities and known colloquially as 'breu', burn brightly under the dampest of conditions and are useful for starting fires or as torches. The Waimiri Atroari class most of the Burseraceae as *wara'* (resin), and in the past employed them for both of the above-mentioned purposes.

Although matches are now used by the Waimiri Atroari for lighting fires, they would traditionally have used a stick and a block of wood, the end of the stick being rotated rapidly in a hole in the block until friction had produced sufficient

* The wood and wood-smoke of some Amazonian tree species are extremely toxic. Certain species of *Ryania* (Flacourtiaceae), for example, to which are attributed numerous bizarre deaths through their accidental use as fuel or grilling spits, contain the highly toxic alkaloid ryanodine (Monachino, 1949).

heat to ignite tinder. This method, or variations upon it, has been (and is some cases still is) used by similar societies throughout the world. The Waiká Indians of northern Amazonia, for instance, employ the wood of *Croton pullei* Lanj. (Euphorbiaceae) in a method almost identical to that used by the Waimiri Atroari (Prance, 1972). At Maré village it was said that in the past they had used the wood of *Guatteria inudata* for the block, which is also favoured for the construction of their palisade walls. Over the species used for the stick, however, there appears to be some confusion. It was said that the stem of *wiriki* was used, which grows in the igapó forest. One informant, when asked to find this species, produced a stick of *Guatteria inundata*, which he also called *ma'duwyxi* but which other people called *amoty*. It was stripped of its bark and would have to be left drying over a fire for a long time before it could effectively be used. However the use of this *Guatteria* has been queried by Grenand; during this study the names *wiriki* and *ma'duwyxi* had also been attributed to certain apocynaceous species (see Table 3), one of which has been recorded elsewhere (French Guiana) as a species used for making fire (Grenand, pers. comm.). There are, however, previous records of the use of Annonaceae for this purpose (e.g. *Anaxagorea* by the Arawak Indians of Corentyn in Guyana (Roth, 1924)), so it is not altogether obvious what conclusion one should reach.

Plants in Ritual and Belief

When Mawa, the first Indian, decided to live in the sky, he summoned the animals to try to shoot an arrow into it. Only the tortoise succeeded. Mawa then sent a woodpecker to tie a vine to the arrow which the tortoise had shot, and climbed it. In his gratitude he invited the tortoise to join him in the sky, but the tortoise lacked the agility to ascend. The spider monkey started to climb, but Mawa, who had not invited him, cut the vine. It fell with the monkey into the tree-tops, and the two have remained there ever since.

The vine that Mawa climbed, *wiamy dydwe* (*Bauhinia guianensis*, Leguminosae), is known colloquially as 'escada de jaboti' or 'tortoise's ladder'. Plants play an important role in Waimiri Atroari ritual and belief, as they do in most tribal cultures, but little of this has yet been recorded. It is said, for instance, that a man who steps on the fruits of *kysiwyra* (*Sloanea* spp., Elaeocarpaceae) will not find tortoises in the forest. There may be complex reasoning behind this superstition, linked to myth or to spiritual concepts, but it could also be interpreted as a metaphorical statement. The fruits of *Sloanea* are covered in spines or irritant hairs, and to step on them with bare feet can be painful. Only an unobservant man will do so, and an unobservant man will not find tortoises as they are notoriously difficult to locate. Thus when it was said that putting the latex of *Clusia insignis* in the eyes will help one to find tortoises, this may be a metaphorical way of saying that it improves or clears the eyesight. However, the Wayãpi Indians of French Guiana, who also use a plant extract (the macerated aerial parts of the orchid *Pelexia goninensis* (Pulle) Schl.) to help them find tortoises, do not put it in their eyes but rub it on their faces (Grenand, 1980). The latter suggests that the correct interpretation may not necessarily be a physiological one.

There are, of course, plant-related beliefs to which no obvious logical explanation is apparent. If a man dreams of snakes, for example, there is a danger that he will be bitten by one in the forest. But if he brushes his legs with a leafy branch of *Helianthostylis sprucei* (Moraceae) before setting out, this will protect him. Similarly a small strip of the fibrous (and strongly smelling) inner bark of *Annona ambotay* (Annonaceae), tied around the ankle of a new born baby, will protect it from the *erkwa* (malevolent spirits). The Kayan people of Sarawak likewise believe that the smell of the inner bark of an annonaceous tree, *Goniothalamus dolichocarpus* Merr., keeps evil spirits at bay (Chai, 1978).

Many plants figure in the *marba* ceremony, keystone of Waimiri Atroari culture, but no detailed account has yet been made. Given the apparent importance of these ceremonies and the possible fragility of the beliefs which support them (when faced with the cynicism of our 'Coca-Cola' culture), it is perhaps desirable that detailed probings are not made. The calendar (*katba*), sent out as an invitation to the participating villages, is made from split sections of *mapyry* (*Gynerium sagittatum*), woven together with *ahnjaty* (*Heteropsis* sp.). Each day one of these sections is twisted by 180° to reveal its split face, until all have been turned and the day of *marba* has arrived. A similar system of invitation is used by the Indians of Guyana (Im Thurn, 1883) whereby knotted strings are distributed, one of which is untied daily.

A simple flute with a haunting note, originally made from a piece of bamboo but recently replaced by PVC piping, is played at certain points in the *marba* festivities, and leafy branches of *Bocageopsis multiflora* (Annonaceae) are used as rattles. At the end of the *marba* these branches are thrown on to the thatch above the door to keep away the *erkwa*. An important part of *marba* is the initiation of young boys, who are whipped soundly with *karwa* (bromeliad fibre) string lashes (*behe*), until their backs are raw and bleeding*. The wounds are washed with an infusion of numerous barks and leaves, after which they are said to heal rapidly (see Plate 6e). A list of some of the component species of this infusion, called *karawyri kany*, is given in Table 2. It appears to have spiritual as well as medicinal significance, since the leaves from which the brew has been prepared are put into palm-leaf packages and suspended above the doors inside the maloca. To lose these bundles was said to be extremely unlucky for the village.

The Waimiri Atroari appear to use neither hallucinogens nor tobacco nor alcohol, although many Amazonian tribes' spiritual/shamanistic practices, festivities or even everyday life involve one or more of these substances (Cooper, 1987). According to Barbosa Rodrigues (1885), however, when he arrived at the village of the Crichaná Indians they offered him, as part of a peace banquet, 'um grande cachiry' (which generally means an alcoholic beverage prepared from manioc). If one assumes that these Indians were the forebears of the present Waimiri Atroari (see discussion on p. 12), it would seem strange that they should have forsaken the practice of preparing alcohol. There are, however, no records by previous workers in the area of its use, although in June 1989 the women of Mynawa were preparing beverages for a *marba* festival (*Mauritia flexuosa* fruits, tapioca starch, extract of *Licania heteromorpha* bark and water) approximately a week in advance of the event.

The Waimiri Atroari do not lack the plants necessary to prepare hallucinogenic drugs; in the hectare of forest which was surveyed there were, for instance, four individuals of four species of *Virola* (Myristicaceae) whose

* A similar practice has been recorded amongst the Macuxi Indians of Roraima (Lowie, 1948).

Table 2
Some ingredients of *karawyri kany*, a spiritually significant 'healing' brew used at *marba* ceremonies

karawyri ia	*Arrabidaea chica* (Bignoniaceae) leaves.
maty bixi	*Qualea* sp. (Vochysiaceae) bark.
kysepe bixi	*Croton lanjouwensis* (Euphorbiaceae) bark.
tykyma wariri	*Piper consanguineum* (Piperaceae).
wiamy iemry ia	*Callichlamys latifolia* (Bignoniaceae) leaves.
dinjeny ia	*Miconia navioensis* (Melastomataceae) leaves.
wiriki ia	*Bonafousia muelleriana* (Apocynaceae) leaves.
kiowyry ia	*Marmaroxylon racemosum* (Leguminosae) leaves.
warma ia	
wa'didimi ia	
kanwe'ra bixi	
ata'myna ia	
clay	
water	

hallucinogenic properties have been documented elsewhere (see Table 3). The resins from the inner bark of these trees contain considerable quantities of indole alkaloids (tryptamines and β-carbolines) (Holmstedt *et al.*, 1980). Curiously, however, in spite of the known psychomimetic effects of these compounds, tests on mice with extracts of the bark resin of *V. elongata* have revealed greater behavioural disturbance from the non-alkaloidal components than from the alkaloidal ones (Macrae & Towers, 1984).

Prance (1972a) reports the occasional use of the yellow powder from the surface of a pyrenocarpous lichen as a snuff by the Dení Indians of Brazil, although it appeared to possess no psychoactive properties. The effects of such a lichen (uncontrollable sneezing) were demonstrated to one of the authors (WM) during the present study by a Waimiri Atroari informant (Warpeki Comprido), but seemingly as a source of amusement rather than as an indication of its use.

Commercial Plants

The Waimiri Atroari, at the time that this study was carried out, were not engaged in any significant commercial activity. Items were occasionally exchanged with Waiwai Indians, and small quantities of craft goods were sporadically sent to Manaus (generally via FUNAI employees), but this had made no substantial contribution to their 'economy'. In some of the villages the Indians have also been encouraged to produce art-work for the employees of the Paranapanema mining group (Baines, 1991). Although a desire for manufactured goods from the outside world was actively being created by the Programa Waimiri Atroari during the course of this study, it was only just reaching the stage where a need for exchangeable commodities is created (cf. Gross *et al.*, 1979). In the past, most of the materials which they have received have been 'given' as compensation for their lost land. Clearly this situation will not continue indefinitely, and since the creation of material desires (which in many cases become material needs) is generally an irreversible step (Colchester, 1989), the Indians are increasingly pressured into entering commerce.

At the village of Maré, in October 1989, the Programa Waimiri Atroari arrangement was such that if a commodity was provided for the entire group (as were hats etc.), then no payment was required, but if an individual requested a specific item then he or she would have to pay for it with craft products. Thus an old man requiring false teeth was requested to make arrows for sale in Manaus to pay for them. To facilitate these activities it was suggested by the local FUNAI representative that the Indians learn to regiment their lives by devoting their Sundays to 'artesanato' (craftwork). Although this type of regimentation is totally inconsistent with their traditional lifestyle, it is not entirely new to them. During the early 1980's the Indians of Maré were required to work an eight hour day in the agricultural plantations which had been set up on their behalf, leaving only the weekends free for hunting etc. (Baines, 1988).

These developments are not discordant with FUNAI's official policy of integration operating in Amazonia at the time of study. Indeed the natural attraction of the Indians to manufactured goods plays a fundamental role in this process, from the initial contacts with isolated tribes (now restricted to those with whom imminent contact is inevitable), to the final 'produção' (production) phase wherein the Indians have acceded, commenced trade, and are regarded as having made a positive step towards becoming members of society. It has been suggested that in preparation for their eventual 'economic independence' (which should in no way be confused with their original true state of independence), the Waimiri Atroari should be encouraged to harvest (and maybe even cultivate) fruit trees such as the Brazil nut *Bertholletia excelsa*. Furthermore, there are factions who believe that having been 'granted' large tracts of land, the Indians bear a responsibility to exploit them for the benefit of Brazilian society.

These are extremely complex and sensitive issues. Although it is currently the subject of much discussion and research, the science of sustainable multi-product rain forest 'extractivism' on a commercial scale is, from a practical point of view, still in its infancy (Fearnside, 1989). Furthermore when such projects are incorporated into the lives of indigenous peoples, the repercussions can be yet more complicated (see Colchester, 1989). Poorly executed projects run a considerable risk of causing irreparable damage to the Indians' sole physical resource, which is their land. There are many documented examples of the occurrence of significant environmental degradation following commercialization (e.g. Gross *et al.*, 1979), and in the worst cases these have ended with the indiscriminate felling and selling of timber trees, generally for a pittance (cf. Greenbaum, 1989; Hildyard, 1989) — or worse still the selling of the land itself. It is impossible to predict exactly how the situation will develop, particularly given the added complication that beneath Waimiri Atroari lands lie valuable and coveted mineral resources, but come what may it seems inevitable that commercial plants will be playing an increasingly important role in Waimiri Atroari life in the not very distant future.

CHANGES IN PLANT RESOURCE UTILIZATION

Maclean (1977), in his discussion of the demise of the St Kildan bird culture, summarized the situation as follows: 'When a simple or primitive society comes into contact with civilization it undergoes cultural change. Unable to resist the seductive advances of a more developed technology, it also embraces or has forced upon it a new system of values which inevitably leads to the breakdown of the old culture. This process of acculturation can be seen today as instrumental not only in the destruction of large numbers of simple societies in under-developed parts of the world, but also as a cause of social and environmental problems in the increasingly homogeneous wasteland of modern civilization'.

Although change is an integral part of any human society, the rate at which the type of change described above takes place almost invariably leaves insufficient time for its accommodation within the original culture. In the case of the Waimiri Atroari these changes have manifested themselves in several ways, not least of which is the way that they regard and utilize the plant resources of their reserve. These changes have been discussed in each of the relevant sections above, but are summarized below:

— Commencement of the use of plant species, resulting from their introduction (e.g. crop plants), from the influence of outsiders (e.g. alternative construction materials) or from the advancement of technology (e.g. acquisition of chain saws for paddle-making etc.).

— Discontinuation of the use of plant species, resulting from substitution by imported items (e.g. modern medicines, sea salt, aluminium canoes), or through the influence of outsiders (e.g. body adornment).

— Alteration of forest management methods, resulting from technological advancement (e.g. chain saws), or the influence of outsiders (e.g. increased sedentariness, cattle ranching).

— Alteration of attitudes towards plant resources, resulting from a combination of factors and possibly symptomatic of other aspects of cultural change.

— Loss of plant resources, resulting from land loss (e.g. to hydroelectric damming).

— Loss of traditional knowledge, resulting from death of elders (e.g. measles epidemics) and the cessation of traditional practices (e.g. medical treatment), combined with disinterest/lack of encouragement of the young and ridicule from outsiders.

Some of these changes are apparently beneficial and logical progressional steps, such as the replacement of clay/ash pots with aluminium saucepans, and the use of higher quality woods for paddle-making since the introduction of chain saws. They have facilitated the traditional way of life without obviously stimulating further change themselves. Others are considerably more complex. The substitution of a traditional technology by another which requires regular input from the outside world (e.g. medicines, shotguns, outboard motors etc.) need not intrinsically be detrimental, but when in the initial enthusiasm for change the traditional technology of the people is allowed to be lost, then the capacity of the community for self-reliance has been weakened. Furthermore if by a 'snow-balling' effect other aspects of technology are allowed to be lost which have not been effectively substituted (see the discussion of dug-out canoes on p. 35), then immediate disadvantages may accrue.

WAIMIRI ATROARI BOTANICAL KNOWLEDGE

The majority of the original information presented in this paper, i.e. that gathered from Waimiri Atroari botanical informants, is traditional tribal knowledge passed down from one generation to the next. However the Waimiri Atroari have not lived in total isolation from other Indian groups, and it is more than likely that some of their technology and knowledge (and indeed language) will have been exchanged or hybridized with that of the other groups with whom they have had contact in the past. Furthermore one cannot discount the possibility that certain details recorded here originate with the FUNAI employees who have been posted in their villages in the more recent past. It is common practice for FUNAI to contract Indians from other groups to assist with the earlier stages of contact with a tribe (particularly if their languages share a common root), and therefore not unlikely that certain of the more impressive elements of those peoples' own traditional knowledge will have been assimilated by their 'hosts'.

The extent to which the Waimiri Atroari's traditional information is being lost, and the ease and rapidity with which such a loss can occur, is discussed to some extent on p. 37 (in the context of medicinal plants). However, certain elements of the information collected during this study may have been specific to the informants who provided it, having stemmed from their individual experience. To what extent this is the case would not be easy to assess, but the fact that on several occasions an informant denied knowledge of the name of a tree whilst stating with great certainty that a particular animal ate its fruits, supports the possibility that individual experience plays a significant part in (at least) this particular aspect of their knowledge.

Individual experience could perhaps be held partially responsible for the differences in opinion which occasionally arose between informants over the uses of certain plant species, but it seems more likely that other factors are accountable. In spite of the remarkable facility for plant identification and taxonomy which these Indians evidently possess, one cannot dismiss the possibility of occasional error. It was clear, for instance, that when dealing with infertile specimens of certain plant families (e.g. Lecythidaceae, Burseraceae, Myrtaceae & Lauraceae), there was a degree of nomenclatural ambiguity. Professional taxonomists will empathize over these difficulties! Given, however, that in many cases the uses attributed to the plants depended upon the names by which they were recognized (see p. 18), those ambiguities must be taken into account when considering the data. In reality the only totally reliable method of collecting ethnobotanical information is by the collection of a plant whose unsolicited utilization has been observed in practice. Regrettably those methods are impractical for a quantitative study such as this, and are rarely employed in modern ethnobotanical research.

The vast majority of the botanical information was gathered from the inhabitants of the village of Maré. On the few occasions when Indians from other villages participated in the study, it became apparent that their opinions and the Maré Indians' opinions on the uses of certain plants were not always the same. Indians from the village of Cacau, for instance, attributed pain-relieving properties to the inner bark of *Croton lanjouwensis* (Euphorbiaceae), although the informants from Maré (at least those present) denied knowledge of such an application. The inhabitants of Maré claimed to use the sap from a forest *Inga* (tree) species to blacken their basket-work, and yet the Indians of Mynawa claimed to use a vine (unobserved), to which they assigned the same name. Similarly Temehe Tomás, the 'captain' of the latter group, denied that *Humirianthera rupestris* (Icacinaceae) is used as a fish poison, although it had been identified as such at Maré by Warpeki Comprido. Furthermore Tomás used several noticeably different pronunciations for the names of common plants. The real or imagined differences between Waimiri and Atroari Indians (see p. 11) will not be dwelt upon further, but nevertheless it seems likely that further studies, in other parts of the Waimiri Atroari reserve, would yield ethnobotanical information significantly different from that presented here.

A list of the plant species collected and observed during this study is presented in Table 3, together with data on habit, habitat, distribution (for most species), common names and uses etc. where available. Species recorded by Grenand *et al.* (in 1987) but not observed during the present study are also included, and are marked GR. Collection numbers of voucher specimens are given in square parentheses. With the exception of the GR specimens, these numbers refer to the Robert Miller (RM) series, collected in 1989. Those few palm species marked SR, for which no collection numbers are available, were recorded by Adauto de Souza Ribeiro as part of Grenand's study. Species marked with an asterisk followed by a number were collected in the hectare quantitative utilization survey. The number refers to the number of individuals with a diameter at breast height >10cm occurring within the plot.

When more than one Waimiri Atroari name has been given for a plant, but the names appear to be alternatives or synonyms, these are listed together. In those cases where the names do not appear to be synonymous but result from differing identifications which have not been resolved, the names and the attributed useful properties are presented separately. Where possible the local vernacular names and the English names are also given, and any additional useful properties gleaned from a scan of some of the available literature. The intention of this is to give a broader picture of the potential usefulness of the species. Furthermore, for the sake of comparison and confirmation, references to other records of medicinal uses and animal relationships which correspond to those assigned by the Waimiri Atroari have also been included. However, references from the literature to congeners of the species collected are generally excluded.

The vernacular names are a mixture of Portuguese and *lingua geral*. The latter is a Tupi-Guarani-based language that was used by Jesuit missionaries in the 17th and 18th centuries for evangelization in Amazonia (Grenand & Ferreira, 1989), and which features strongly in contemporary nomenclature of the natural world. Some vernacular names are borrowed from other Indian languages, including those of the Carib group.

References to the literature are numbered and presented in parentheses. The numbers refer to those in the reference list on pp 122–136.

Table 3
Plant species recorded in the Waimiri Atroari reserve

Note: For some of the tree species recorded within the hectare plot, many of which were in a sterile state at the time of collection, there was a degree of doubt or dispute amongst the informants over the identification (i.e. Waimiri Atroari name) and/or the useful properties of the species (see discussion on p. 47). In the list that follows, the scientific names of those species are preceded by a hyphen.

For an explanation of the format of this table, and of the codes employed (*, GR, etc.), see previous page.

ANACARDIACEAE

Anacardium occidentale L.

[no voucher taken]

Tree cultivated around villages.

Widely cultivated throughout the tropics, native to Atlantic coastal South America.

hiri (Waimiri Atroari). The fleshy pedicel is edible.

caju (local vernacular).

cashew (English). An alcoholic beverage is made from the fruit of this tree in northeast Brazil (113). The Tikuna Indians of Colombia drink the juice of the pedicel to cure influenza (257). The seed is edible and nutritious when roasted. A resinous oil from the pericarp, known as cardol, is used by the Wayãpi Indians of French Guiana to cauterize splits in the soles of the feet (122). This oil is employed in the varnish, plastics and insecticide industries (62), in the manufacture of brake-blocks (170), and has been used to treat malaria, toothache and syphilitic ulcers (87). The gum exuded from the bark has been used as a substitute for gum-arabic for book-binding (62), as indelible ink, as a cold cure (in Cuba), and is attributed fungicidal and insect-repellent properties (87). A decoction of the bark is held to act as a contraceptive in Colombia (257) and as a cure for sore throat and tumours in Brazil (62). It is used in French Guiana as a diarrhoea treatment by the Créoles and a cure for infantile thrush by the Wayãpi Indians (122), and in the Distrito Federal of Brazil as a remedy for liver disorders (115). In Java and Malaya the leaves are eaten raw with rice (87). This species contains alkaloids (257), and possesses anti-bacterial properties (281). The multifarious uses made of this plant in Africa, which include preservation of fishing nets with cardol in Zanzibar and use of the bark as a fish-poison in the Congo, are summarized by Watt & Breyer-Brandwijk (1962).

Mangifera indica L.

[no voucher taken]

Tree cultivated around villages.

Cultivated throughout the tropics, native of Asia.

manga (Waimiri Atroari; borrowed from vernacular). The fruits are edible.

manga (local vernacular).

mango (English). This species serves as an urban shade-tree, the seed is a source of flour and the wood may be used for floor boards, tea-chests etc. (170). The young leaves may be eaten, either raw or cooked (280). The Tikuna Indians of Colombia use a decoction of the leaves as a contraceptive or abortifacient (257). The Chácobo Indians of Bolivia drink a similar decoction for rheumatism (48), and the Créoles of French Guiana take it for diarrhoea and stomach ache (122). The Palikur Indians of French Guiana treat prickly heat (of babies) with plasters of the mashed leaves (which contain tannins) (122), and the Créoles of the Antilles drink a decoction of the flowers for angina and asthma (191). The leaves also contain saponins, glycosides, unsaturated sterols and polyphenols, and are used by the Warao Indians to treat fevers, coughs and diarrhoea (287). They possess antibiotic properties (280).

Thyrsodium paraense Huber GR
[GR2751]
Tree in *terra firme* forest.
kanabowa (Waimiri Atroari). The trunks are used for house-posts.
breu de leite (local vernacular; = milk resin). This species is attributed (unspecified) medicinal properties by the Ka'apor Indians of Brazil (23). The Wayãpi Indians of French Guiana use a closely related species (*T. spruceanum* Benth.) for their house-posts (Grenand, pers. comm.).

ANNONACEAE

Annona ambotay Aublet
[no voucher retained]
Tree in *terra firme* forest.
Widespread in Amazonia, extending north into the Guianas (48).
karaxmia (Waimiri Atroari). To cure 'madness' the forehead is perforated with a spine or piece of glass, and a strip of the fibrous inner bark tied around the head. Alternatively, the bark can be burned and the smoke inhaled. A small strip of bark is tied around the ankle of a new-born baby to ward off malevolent spirits (*erkwa*).
envirataia (local vernacular; *envira* = fibre, *taia* = peppery). The Chácobo Indians of Colombia drink a decoction of the leaves for headache, apply scrapings of the bark to skin sores, and eat the fruits (48). In the region of Alter do Chão, Pará, Brazil, an infusion of the bark is drunk for stomach ache, and the smoke of its burning wood is used to cure various ailments (50). The Wayãpi Indians of French Guiana use the leaves or bark to prepare a febrifuge, and the Palikur use a decoction of the leaves as a wash in childbirth (122). The Tecuna (Tikuna) Indians of Colombia include this species in their curare arrow poison (8) and the Galibis of French Guiana use a decoction of the bark to cure ulcers (208). Four major alkaloids have been isolated from the plant, whose effects are analgesic, antibacterial and antispasmodic (135 *fide* 122).

Annona montana Macfad.
[no voucher taken]
Tree cultivated around villages.
Native of Amazonia and widely cultivated throughout.
araxky, **araxki** (Waimiri Atroari; possibly borrowed from local vernacular). The fruits are edible.
aráticum, **aráticum cagão** (local vernacular).
mountain soursop (English). A decoction of the leaves is taken as a sedative by the Créoles of French Guiana (122). This species contains cytotoxic compounds (192).

Annona muricata L.
[no voucher taken]
Tree cultivated around villages.
Widely cultivated throughout the tropics, native of tropical America (Antilles?).
araxky, araxki (Waimiri Atroari). The fruits are edible.
graviola (local vernacular).
soursop (English). An infusion of the leaves may be taken to lower blood sugar levels (62), and is taken as a sedative by the Créoles of French Guiana (122). The Island

Caribs of Dominica believe that the fruits, eaten or used as a poultice, stimulate lactation in women (136). In Jamaica the fruits and leaves are held to be anthelmintic (292), and toxic compounds have been reported from the seeds (54). Extracts of the leaves have a powerful hypotensive effect on rats (58). In Africa the root bark, which contains hydrocyanic acid, is used as a fish poison (285).

Bocageopsis multiflora (Mart.) R.E. Fries *6
[597] [GR2739]
Tree in *terra firme* forest.
Northern Brazilian Amazonia and Guyana (105).
iakba, iakpa (Waimiri Atroari). The leafy branches are used as rattles at *marba* ceremonies, and subsequently thrown onto the thatch above the maloca doors to ward off spirits. The wood is used for house construction. The fruits are eaten by curassows, toucans, guans and howler monkeys.
envira surucucu (local vernacular; *envira* = fibre, *surucucu* = bushmaster snake).

Duguetia cauliflora R.E. Fries *1
[393]
Tree in *terra firme* forest.
Guyana, Surinam (105) and Brazilian Amazonia.
xaxerany, xaxirany (Waimiri Atroari). Scrapings of the inner bark, or pieces of the fruits, may be chewed to relieve toothache. The trunks are used for roof beams.
xaxerany (Waimiri Atroari). Hunting dogs are made to sniff pieces of the cut bark to improve their performance. The trunks are used for roof beams. The fruits are eaten by capuchin monkeys.

Duguetia flagellaris Huber
[295] [GR2713]
Slender treelet common in *terra firme* forest.
Northern Brazilian Amazonia.
sasra, sasry (Waimiri Atroari). The slender trunks are used to make fishing spears, fishing rods (*fide* Grenand) and thatch supports (*iara bahnja*) in house construction. An infusion of the bitter-tasting bark is used as a cough and cold medicine and as *ka'na ieptxi* ('mosquito medicine').
envira amarela (local vernacular; = yellow fibre). An extract of the bark is used to cure rheumatic pains in Peru (9), and contains alkaloids (232).

Duguetia sp. nov. *3
[199]
Tree in *terra firme* forest.
xaxerany (Waimiri Atroari). The trunks are used in house construction. The fruits are eaten by capuchin monkeys.

—***Ephedranthus amazonicus*** R.E. Fries *2
[434]
Tree in *terra firme* forest.
Central Brazilian Amazonia (vicinity of Manaus) (105).
pakia de'de (Waimiri Atroari; *pakia* = collared peccary, *de'de* = forehead). The wood is used to make clubs for killing peccaries (see p. 31). Strips of the fibrous inner bark are used as carrying straps. The trunks are used in house construction. The fruits are eaten by capuchin monkeys.
amoty (Waimiri Atroari; = fibre). The fruits are said to be edible (although this has been questioned by Grenand). Strips of the fibrous inner bark are used as carrying straps.
envira dura (local vernacular; = tough fibre).

—*Guatteria inundata* Mart. ex R.E. Fries
[791]
Treelet common on the margin of the igapó forest.
Brazilian Amazonia (105).
amoty (Waimiri Atroari; = fibre).
wiriki, ma'duwyxi (Waimiri Atroari). In the past the slender stems were stripped, thoroughly dried, and used as fire-sticks by rotating the point rapidly in a block of *Guatteria olivacea* wood. There may, however, be some confusion here. The name *wiriki*, which refers to the plant used for fire-making, was also applied to two apocynaceous shrubs and the name *ma'duwyxi* to another (see discussion of *Malouetia furfuracea*). The word *ma'duwyxi* apparently refers specifically to the wood of this plant, which is used for fishing reels (Baines, pers. comm.).

This species contains alkaloids in low concentrations (232).

Guatteria olivacea R.E. Fries
[792]
Tree in *terra firme* forest.
Brazilian Amazonia (105).
amoty (Waimiri Atroari; = fibre). The trunks are split into stakes which are used for the palisade walls of the house (said to be particularly resistant to spear attack). The wood was used for fire-making in the past (see *G. inundata)*. Fibres from the bark may be used as carrying straps.

Rollinia insignis R.E. Fries var. ***pallida*** R.E. Fries *1
[446]
Tree in *terra firme* forest.
Peru (105) and Brazilian Amazonia.
amoty (Waimiri Atroari; = fibre). Strips of the fibrous inner bark are used as carrying straps. The wood is good for fuel.
envira bobó (local vernacular; *bobó* = clown or a type of Bahian food). The wood is used for construction etc. in Brazilian Amazonia (166).

Rollinia mucosa (Jacq.) Baill.
[no voucher taken]
Tree cultivated around houses.
Cultivated in northern South America and the Antilles, native of Amazonia.
briba (Waimiri Atroari; *fide* Grenand; presumably borrowed from the vernacular). The fruits are edible.
biriba (local vernacular).
queen fruit (English). The Wayãpi Indians of French Guiana take the seeds of this tree from wild stock and plant them in their swidden clearings (120). The wood may be used to make canoes and boat masts, and the powdered seeds may be taken as a cure for enterocolitis (280).

Unonopsis duckei R.E. Fries *1
[754]
Tree in *terra firme* forest.
Brazilian Amazonia (105).
pakia de'de (Waimiri Atroari; *pakia* = collared peccary, *de'de* = forehead). The wood is used to make clubs for killing peccaries. The fibrous inner bark is used for carrying straps. The trunks are used in house construction. The fruits are eaten by capuchin monkeys

This species contains alkaloids (232). A closely related species, *U. perottetii* Fries, is regarded as the best wood for construction by the Wayãpi Indians of French Guiana (120).

Xylopia calophylla R.E. Fries *1
[493]
Tree in *terra firme* forest.
Brazilian Amazonia and Colombia (105).
txiri (Waimiri Atroari). The trunks are used for house construction. The fruits are eaten by toucans and toucanets.

APOCYNACEAE

Aspidosperma nitidum Benth.
[789]
Tree in *terra firme* forest.
Upper Amazon valley (290).
pamda (Waimiri Atroari). Small pieces of the bitter-tasting bark may be chewed to cure toothache. Axe handles are made from the fluted trunk.
carapanaúba (local vernacular; = mosquito tree in Tupi). The bark is much used in local folk medicine for malaria and 'liver' problems. In the State of Roraima an infusion of it is taken for inflammation of the ovaries and uterus (41), and in the State of Amazonas as a cure for diabetes, stomach problems and cancer (40). It contains indolic alkaloids of the corynantheol type (16). The Makuna and Taiwano Indians of Colombia use the latex to cure leprosy, which is prevalent in the region (250). In the Guianas the inner bark of *Aspidosperma excelsum* Benth. is also used to treat toothache (97).

Bonafousia muelleriana (Mart.) Boit. & L. Allorge (= *Tabernaemontana macrocalyx* M. Arg.)
[798]
Shrub/treelet in *terra firme* forest.
Widespread throughout Amazonia (6).
wiriki (Waimiri Atroari). The leaves are an ingredient of the *karawyri kany* healing brew used at *marba* ceremonies (see p. 43).

The fruits have been recorded as edible in Loreto, Peru, but in the same country it has been noted that the fruits and roots are toxic and used to poison arrows (6). A closely related species, *B. morettii* L. Allorge, is also said by the Wayãpi Indians of French Guiana to be poisonous, and is used to repel evil spirits (120).

Bonafousia undulata (Vahl) A. DC.
[773]
Small treelet in *terra firme* forest.
Widespread throughout Amazonia and the Guianas, and in Venezuela, the Caribbean and Central America (6).
wiriki (Waimiri Atroari). Scrapings of the bark are mixed up with a little water or spittle and applied to fungal infections (& mite infestations ?) of the skin.

An extract of the bark of *B. undulata*, which is rich in indolic alkaloids, is used by the Palikur Indians of French Guiana to treat swollen feet. The grated bark is also applied as a compress to treat headache (122). The Taiwano of the Kananarí River (Colombia) use a decoction of the leaves, mixed with manioc leaves, as a vermifuge (250). In Guyana the latex is used to remove botfly larvae from the skin (97).

Couma macrocarpa Barb. Rodr. *4
[259]
Tree in *terra firme* forest.
Central America, Venezuela, the Guianas and Amazonia (182).
kamnjeni, kamnjany (Waimiri Atroari). The fruits are edible and sweet. The latex is sweet and may be drunk in small quantities, or mixed with tapioca. Latex is tapped from the trunk, boiled until glutinous and mixed with the latex of *Symphonia globulifera* (Guttiferae) to make the blocks of *manji* resin used as glue for arrow-making etc. (see p. 29). Wood from the upper parts of the tree is used to make the benches (*miri*) which are taken to *marba* ceremonies. The fruits are eaten by monkeys.
sorva (local vernacular). The latex has been exploited commercially for the production of chewing gum (231). The Andoque Indians of Colombia drink it as a cure for diarrhoea (162), and the Indians of the Vaupés River of Colombia use it to heal the umbilicus of the new-born (257). It is said to be a cure for intestinal amoebiasis (198). The Witoto Indians of Peru chew the leaves of this species as a substitute for coca, although no psychoactive constituent has yet been identified (257). The wood is used for construction and carpentry in Brazilian Amazonia (166). The fruits are eaten by woolly monkeys (144).

Lacmellea ramosissima (Muell. Arg.) Markgraf *1
[97]
Tree in *terra firme* forest.
Northern Amazonian Brazil (176).
kiawyxi, iai (Waimiri Atroari). The fruits are edible (sweet). The latex is sweet and may be drunk in small quantities. The fruits are eaten by monkeys and by tayras.

Malouetia furfuracea Spruce GR
[GR2758]
Shrub or small tree in *terra firme* or flooded forest.
Brazilian Amazonia and the Guianas (289).
ma'duwyxi (Waimiri Atroari). There may be some confusion between this species and *Guatteria inundata* over their traditional uses as fire-sticks. The Palikur Indians of French Guiana used *M. tamaquarina* A. DC. for that purpose (Grenand, pers. comm.), which is considered by some to be synonymous with *M. furfuracea* — and whose wood is good for making small utensils such as spoons (289).

Parahancornia amapa (Hub.) Ducke
[no voucher retained]
Tree in *terra firme* forest.
Northern Brazilian Amazonia and the Guianas (182).
amypa (Waimiri Atroari). The fruits are edible and much sought after.
amapá amargoso (local vernacular; *amargoso* = bitter). The latex is taken (with milk) by the Palikur Indians of French Guiana to treat internal ulcers, and the roots (and to a lesser extent the leaves) contain tannins and saponins (122). This species is used to cure stomach ache by the Ka'apor Indians of Brazil (23). The Tembé Indians of Brazil drink the latex as a cure for hepatitis (24). The latex may also be taken as a cure for syphilis or as a tonic (39), and the *caboclos* (mixed-blood country people) of the lower Amazon drink it to cure bronchitis (10). In Peru the latex is used to seal cuts and is sometimes sold as chicle, mixed with the latex of *Couma macrocarpa*. In Brazil it is applied as a poultice to contusions (163). The wood is used in the vicinity of Manaus for construction (261), and is an excellent substitute for white pine (96). The Wayãpi Indians of French Guiana use it to make mediocre dug-out canoes (120).

ARACEAE

Dieffenbachia sp.
[779]
Herb in regrowth alongside track in *terra firme* forest.
myky (Waimiri Atroari). A small piece of the medulla of the stem is inserted into the cavity of an aching tooth with a fine splinter of wood, and dulls the pain. This plant was said to be poisonous.

Dieffenbachia species, which are generally recognized as poisonous on account of the abundant calcium oxalate crystals which are contained in their tissues (together with several other toxic compounds), have been reported to be used as arrow poisons, sterility drugs and cures for gout, dropsy, impotence and frigidity (15), and also as a cure for leishmaniasis (122). The use of *Dieffenbachia* as a pain-killer has also been reported from the Cayapa and Quichua Indians of Ecuador, where *D. seguine* (Schott) Baill. is used to treat the stings of ponerine ants (157).

Heteropsis sp. (probably *H. jenmani* Oliver)
[no voucher taken]
Climbing epiphyte very common in *terra firme* forest
ahnjaty, axnjaty (Waimiri Atroari). The flexible aerial roots are used for lashing beams, attaching thatch etc. in house construction, and as a general binding/tying material. Stripped of their bark and split, they are also used to lash together the component parts of *wiepi* panniers, and to make *katba* calendars.
cipó titica (local vernacular; *cipó* = vine).
This plant is used as the main element of basketry by many Amazonian tribes.

Philodendron goeldii G.M. Barroso GR
[no voucher taken]
Climbing epiphyte in *terra firme* forest.
Widespread throughout Amazonia and the Guianas (178).
weme (Waimiri Atroari). The aerial roots are used as a tying and binding material.
cipó ambé (local vernacular).

Genus undetermined
[no voucher taken]
Climbing epiphyte in *terra firme* forest.
ximkoby (Waimiri Atroari). The sap from the aerial roots may be put into the eyes to 'find tortoises in the forest' (see p. 42). Scrapings of the bark are rubbed on the stings of ponerine ants to ease the pain.

The sap of an unidentified Araceae is used by the Paumarí Indians of Brazil to treat various Hymenopteran stings (223).

BIGNONIACEAE

Arrabidaea chica (Humb. & Bonpl.) Verl. GR
[GR2753]
Scrambling vine cultivated in swidden clearings.
Mexico to Argentina (109).
karauri (Waimiri Atroari). The dried leaves are employed in initiation rituals.
crajiru (local vernacular). An infusion of the leaves is drunk by the *caboclos* of the lower Amazon to treat menstrual and intestinal problems and hepatitis (10). In the State of Amazonas it is taken for 'blood' and 'nerve' problems and used to bathe wounds (40).

It is also reputed to relieve skin inflammations (261). The Tikuna Indians of Colombia use an infusion of the leaves as an eye-bath to treat conjunctivitis (257), and the Jívaro Indian women of Peru chew the leaves to blacken their teeth and prevent caries (164). The leaves produce a red dye which is used by many tribes of Amazonian Indians, and it has been reported that in the region of the Orinoco they are fermented to produce an alcoholic beverage (53).

Callichlamys latifolia (Rich.) K. Schum.
[778] [GR2711]
Single-stem shrub or woody liana in *terra firme* forest.
Southern Mexico to Brazil and Ecuador (109).
wiamy iemry, (Waimiri Atroari; = tortoise's penis). An infusion of the bitter-tasting bark is drunk to relieve fever. According to Grenand *et al.* (pers. comm.) this species is used to prepare a wash employed during childbirth.

The Wayãpi Indians of French Guiana use a decoction of the bark to treat leishmaniasis. The roots contain high concentrations of naphthoquinones (122).

Crescentia cujete L.
[no voucher taken]
Tree cultivated around villages.
Cultivated throughout the tropics, but native to the New World.
bixiwa (Waimiri Atroari). The fruits are used as drink and food containers.
cuieira (local vernacular).
calabash (English). In southern Peru the leaves are chewed to relieve toothache (257). The young fruits may be pickled and eaten (170). The seeds may be roasted and eaten, and the Chiricano Indians of Panama drink an infusion of the pulp of the unripe fruits to alleviate the rigours of childbirth (87). This pulp, believed to be poisonous by the Wayãpi of French Guiana, forms the basis of a contraceptive decoction used by the Créoles of that country (122). In Jamaica it is heated and used as a poultice for bruises and inflammations (292), and in the Guianas it is sometimes used as fish bait (239). The Créoles also use an infusion of the leaves as a purgative, and take the juice of the young fruits to cure diarrhoea (122). This species contains compounds with antibacterial properties (281).

Jacaranda copaia (Aublet) D. Don *1
[638]
Tree in *terra firme* forest.
Throughout tropical South and Central America (48).
wa'by (Waimiri Atroari). The wood may be used in house construction (in the roof), and is good for fuel. The fruits are eaten by macaws and parrots.
caroba (local vernacular). The wood is used to make matches, paper etc. in Brazil (166). A decoction of the leaves is drunk by the Chácobo Indians of Bolivia to treat rheumatism (48). The Tiriyó of northern Brazil bathe in an infusion of the leaves to cure fever and general weakness (63), and in French Guiana the bark of young individuals is used to prepare an emetic/purgative, and a cure for syphilis (128). The leaves are burned as a mosquito repellent in the region of Alter do Chão, Pará, Brazil (50), and as a blackfly repellent by the Wayãpi Indians of French Guiana (122). The Andoque Indians of Colombia mash the leaves and boil them in water to the consistency of honey, and apply the mixture to wounds to assist healing (162). In French Guiana the leaf sap is used by the Negroes to treat leishmaniasis (208). Indians of the Vaupés River of Colombia employ the sap of the bark for curing skin infections (257), and the Jívaro of Peru crush the leaves and apply them directly for the same purpose (164). This species is used as a fish poison in the Guianas (151).

According to the Wayãpi Indians of French Guiana, jaguars poison their claws with the sap (120). The flowers are eaten by tortoises (188).

Tynanthus cf. ***polyanthus*** (Bur.) Sandw.
[GR2787]
Woody liana in *terra firme* forest.
watypyra (Waimiri Atroari; *waty* = fire). An infusion of the bark, which smells strongly of cloves, may be drunk as a cure for colds (and for pleasure?). According to Grenand *et al.* (pers. comm.), this is used to flavour a drink made from bananas.
cipó cravo (local vernacular; = clove vine). The root bark of *T. elegans* Miers is used as a stimulant and aphrodisiac in the vicinity of Manaus (261).

—Genus undetermined
[794]
Scrambling vine in secondary vegetation.
karawyri, karewyri (Waimiri Atroari). The leaves of this vine are the principal ingredient of the *karawyri kany* healing brew used at *marba* ceremonies (see p. 43). **Note:** Grenand *et al* collected a specimen [2753] of *Arrabidaea chica* under the same Waimiri Atroari name, and it is probable that the Miller collection [794] was incorrectly identified as *karawyri* by the informant (Baines (pers. comm.) records a similar mistake). This likelihood is supported by the fact that Im Thurn (1883) collected *A. chica* amongst the Carib Indians of Guyana under the similar name of *caraweera*, where it was used as a dye. There is considerable leaf heteromorphy in this species (Grenand, pers. comm.).

BIXACEAE

Bixa orellana L.
[no voucher taken]
Shrub cultivated around villages.
Native to and widespread throughout the Neotropics.
kiswe, kyswi (Waimiri Atroari). A red dye obtained from the aril of the seed is sometimes used for body painting.
urucum (local vernacular).
anatto (English). The aril is used throughout the world as a food dye, and is sometimes used for colouring basketry etc. The Chácobo Indians of Bolivia eat the seeds, cooked in butter (48). In French Guiana these seeds, together with the pulp, are used as a vermifuge by the Wayãpi Indians (who also treat abcesses with the juice squeezed from the heated buds). The Créoles of that country use an extract of the petioles to clear the eyes of encrustations due to blepharitis (122). In Colombia the red dye, if ingested, is said to be an effective antidote for cyanide poisoning (e.g. from untreated manioc) (257), and an infusion of the entire plant is used for the same purpose by the Créoles of the Antilles (191), who also employ the seeds as a febrifuge. The aril, smeared over the body, is commonly used as an insect repellent (J.A. Ratter, pers. comm.). In the Distrito Federal of Brazil the pulp is taken (either with water or honey) to cure asthma, measles and bronchitis (115).

BOMBACACEAE

Eriotheca (Bombax) sp. *1
[363]
Tree in *terra firme* forest.

kirikiri (Waimiri Atroari). Strips of the fibrous inner bark of young trees are used for carrying game. The fruits are eaten by pacas.

Quararibea ochrocalyx (K. Schum.) Fischer *2
[737]
Tree in *terra firme* forest.
Amazonia (Peru and Brazil) and the Guianas (91).
amakra i'riba tany, wirxi (Waimiri Atroari; *amakra* = tucunaré fish, *i'ri* = fruit tree, *tany* = place). The fruits are edible, and are eaten by howler and capuchin monkeys.

In Peru the wood is used in light construction (202). The fruits are eaten by capuchin monkeys (202).

Scleronema micranthum Ducke
[GR2736]
Tree in *terra firme* forest.
waiwane, (Waimiri Atroari). The bark may be removed entire, flattened, and used to make the walls of houses (the use of bark for walls was also mentioned by Barbosa Rodrigues (1885) in his observations on the Crichaná Indians). Strips of the fibrous inner bark are used as carrying straps. The trunk (?) may be used for canoe construction. The wood is good for fuel.
cardeiro (local vernacular). The wood is used in the vicinity of Manaus for furniture and construction (195), and possesses particularly good elastic properties (233).

BORAGINACEAE

Cordia sprucei Mez GR
[GR2754]
Shrub in secondary vegetation.
ieri (Waimiri Atroari; apparently the generic name for fruits with little pulp). The fruits are edible.

BROMELIACEAE

Ananas comosus (L.) Merr.
[no voucher taken]
Herb cultivated around villages and in swidden clearings.
Cultivated throughout the tropics, native of South America.
nana (Waimiri Atroari; possibly derived from another language). The fruits are edible.
abacaxi (local vernacular).
pineapple (English). An alcoholic beverage is prepared from the fruits by certain Brazilian Indian peoples (113). The leaves furnish a source of useful fibre, which is used in the Philippines to produce a fabric (137). The juice of the fruits is a diuretic (87). The Island Caribs of Dominica use the juice of the immature fruits as an abortifacient (136), as do the Tikuna Indians of Colombia (257). In Brazil this species has been used for pus extraction, curing diphtheria and throat disorders, and as a digestive (74).

Bromelia sp. (probably *B. karatas* L. or *B. pinguin* L.)
[no voucher taken]
Terrestrial bromeliad cultivated in swidden clearings.
karwa (Waimiri Atroari; probably borrowed from Tupi-Guarani). Fibres from the leaves are used to make a strong cord which is used for a multitude of purposes, but

primarily for stringing hunting bows. A leaf of this plant may be wrapped around the wrist as a bracer, to protect it from the bow-string. The fibres are also used for fletching arrows.

caraua (local vernacular).

caraguata (English). The fruits of *B. karatas* are eaten by the Wayãpi Indians of French Guiana (120).

BURSERACEAE

The majority of the Burseraceae collected during this survey were identified as either *aiywa* or as *wara'*, and were attributed the same useful properties. In several cases they were attributed both of these names (by different informants). Use of the name *aiywa* for Burseraceae (or similar forms such as 'ayawa', 'hayawa') is common amongst the Arawak and Carib languages in the Guianas, and has also been recorded in Tupi languages (e.g. Wayãpi). Similar names to *wara'* are also encountered in those language groups (Grenand, pers. comm.). The local vernacular name 'breu' (resin) is likewise applied to the majority of Burseraceae.

Resin exuded from the bark (*wara'*) is highly flammable and may be used for lighting fires or for torches (although more rarely since the introduction of electric flashlights). To collect the fine soot used to dye the strips of *Ischnosiphon* (Marantaceae) used for decorative basketry, pieces of the resin are burned beneath a pot (see p. 32). The wood is good for fuel. The fruits are eaten by guans and toucans.

Burseraceous resins are used in the preparation of fine body paints by the Wayãpi Indians of French Guiana (120).

Note: The quantity and quality of resin exuded from a tree will vary considerably from species to species, and probably depends also upon the tree's age and size. Thus although no nomenclatural distinctions may have been made between them, certain species or individuals may not actually be used for these purposes. A detailed study has been made of the classification of these tree resins used by the Ka'apor Indians of the eastern Amazon (28).

Dacryodes nitens Cuatrec. *3

[81]

Tree in *terra firme* forest.

aiywa, wara' (Waimiri Atroari). See above for uses etc.

According to the Wayãpi Indians of French Guiana, this species yields very little resin (120).

Dacryodes roraimensis Cuatrec. *2

[674]

Tree in *terra firme* forest.

aiywa, wara' (Waimiri Atroari). See above for uses etc.

Dacryodes sp. *2

[212]

Tree in *terra firme* forest.

wara' (Waimiri Atroari). See above for uses etc.

Protium apiculatum Swart *27

[348]

Tree common in *terra firme* forest.

aiywa, wara' (Waimiri Atroari). See above for uses etc.

—*Protium decandrum* (Aublet) March. *3
[767]
Tree in *terra firme* forest.
Widespread throughout Amazonia and the Guianas (79).
wara', sakyra iepkry (Waimiri Atroari; *sakra* = white, *iepkry* = sap). See above for uses etc.
tynabwa sakra (Waimiri Atroari). The trunks are used for house-posts. The fruits are eaten by curassows.
breu branco (local vernacular). The Ka'apor Indians of Brazil attribute medicinal properties (not defined) to this species (23), and eat its fruits. The fruits are also considered edible by the Tembé Indians of Brazil (24), and the wood is used to make furniture in Guyana (96). The resin is potentially useful for fumigants and marking ink (97), and in French Guiana is used as incense (208). The fruits are eaten by parrots and brocket deer (28).

Protium fimbriatum Swart *1
[312]
Tree in *terra firme* forest.
wara', minja iepkry (Waimiri Atroari; *minja* = tapioca, *iepkry* = sap). See above for uses etc.

Protium hebetatum Daly *37
[675]
Tree common in *terra firme* forest.
aiywa, wara' (Waimiri Atroari). See above for uses etc.

Protium nitidifolium (Cuatrec.) Daly *3
[463]
Tree in *terra firme* forest.
wara' (Waimiri Atroari). See above for uses etc.

Protium cf. ***opacum*** Swart ssp. ***opacum*** *1
[476]
Tree in *terra firme* forest.
Widespread throughout most of Amazonia (79).
wara' (Waimiri Atroari). See above for uses etc.

In Loreto, Peru, this species is attributed minor commercial value (204). It contains propacin, a coumarinolignoid (296).

Protium paniculatum Engl. var. ***riedelianum*** (Engl.) Engl. *1
[190]
Tree in *terra firme* forest.
Amazonian Brazil (Pará and Amazonas), common around Manaus (79).
wara' (Waimiri Atroari). See above for uses etc.

—*Protium polybotryum* (Turcz.) Engl. ssp. ***blackii*** (Swart) Daly *1
[200]
Tree in *terra firme* forest.
Amazonian Brazil (southern Pará, Amazonas, Rondônia and Mato Grosso States) (79).
kaparwa (Waimiri Atroari). The fruits are edible, and are eaten by guans, toucans, and by capuchin and spider monkeys. **Note**: It is possible that this species was misidentified in the field, since other ethnobotanical studies in which it has been collected have made no reference to the edibility of the fruits.

The fruits are eaten by spider monkeys (238).

Protium* cf. *sagotianum Marchal *1
[535]
Tree in *terra firme* forest.
Widespread throughout Amazonia and the Guianas (79).
wara' (Waimiri Atroari). See above for uses etc.

P. sagotianum is attributed medicinal properties by the Ka'apor Indians of Brazil, who also eat the fruits (23). The Wayãpi Indians of French Guiana likewise eat the fruits (120). In Guyana the wood is used for furniture (96). The resin is a substitute for incense gum, and can be used in the manufacture of fumigants and printing ink (97). The fruits are eaten by parrots (28).

Protium spruceanum (Benth.) Engl. *1
[334]
Tree in *terra firme* forest.
Widespread in Amazonia, central Brazil and inland tropical southern Brazil (79).
xipi iepkry (Waimiri Atroari; *iepkry* = sap). The resin exuded from the bark is eaten or rubbed on the stomach to cure stomach ache. The fruits are eaten by toucans.

In the vicinity of Alter do Chão, Pará, Brazil, an infusion of the leaves is used to cure fever, and the smoke of the burning resin is inhaled to relieve headaches (50). The Ka'apor Indians of Brazil use the resin to relieve toothache, and add shavings of the stem to tobacco to sweeten the taste of the smoke. They also eat the fruits (28). The fruits are eaten by tortoises (28).

Protium subserratum (Engl.) Engl. *2
[311]
Tree in *terra firme* forest.
Widespread throughout Amazonia; also Guyana, French Guiana and Antioquia (Colombia) (79).
kaparwa (Waimiri Atroari). The fruits are edible, and are eaten by guans, toucans, and by howler and spider monkeys.
wara' (Waimiri Atroari). See above for uses etc.

—*Protium urophyllidium* Daly sp. nov. ined. *4
[407]
Tree in *terra firme* forest.
aiywa, wara' (Waimiri Atroari). See above for uses etc.
tynabywa (Waimiri Atroari — a name also given for *Thyrsodium* (Anacardiaceae)). The trunks are used for house-posts. The fruits are eaten by curassows.

Protium sp. *3
[82]
Tree in *terra firme* forest.
aiywa, wara' (Waimiri Atroari). See above for uses etc.

Tetragastris panamensis (Engl.) O. Kuntze *2
[299]
Tree in *terra firme* forest.
Central America and the Guianas (270).
kaparwa (Waimiri Atroari). The fruits are edible, and are eaten by guans, toucans, and by capuchin and spider monkeys.
wara' (Waimiri Atroari). See above for uses etc.

breu preto (local vernacular; = black resin). The wood is used for furniture and cabinet-making in Guyana (96). The fruits are eaten by spider monkeys (238) and by howler monkeys (180).

Trattinnickia glaziovii Swart *2
[426]
Tree in *terra firme* forest.
Brazilian Amazonia (270).
xipi, xipi iepkry (Waimiri Atroari; *iepkry* = sap). Resin exuded from the bark is eaten or rubbed on the stomach to cure stomach ache. The fruits are eaten by toucans.

CARICACEAE

Carica papaya L.
[no voucher taken]
Tree cultivated in swidden clearings and around villages.
Cultivated throughout the tropics, native of the New World.
mabaia (Waimiri Atroari). The fruits are edible.
mamão (local vernacular; = large breast).
pawpaw (English) The young negroes of the Antilles used to make musical instruments from the hollow petioles (124). The seeds are widely used as a vermifuge, and the milky juice — and leaves — as a meat tenderizer (due to the presence of papain, a proteolytic enzyme). Papain is also used for fining beer and shrink-proofing wool (170). In southern India the seeds are taken to induce abortion (285). Latex from the unripe fruits possesses fungicidal properties, is used to treat ringworm in Papua New Guinea (288), and is applied to infected sores by the Aligandi Cuna Indians of Panama (87). In Belize it is applied to wounds (226). In Samoa the inner bark is used to treat toothache (276). The Tikuna Indians of Colombia eat the immature fruits with aspirin to induce abortion (257), whereas in New Guinea the stem latex is ingested for the same purpose (288). The leaves (which are boiled and bound over ulcers in New Guinea (87), are used as a tobacco substitute in New Caledonia and as a soap substitute in Africa (285)), contain carpaine - an alkaloid which depresses the central nervous system and acts as a potent amoebicide (133). In Belize they are used as a purgative for sick horses (226), and the Wayãpi Indians of French Guiana add them to their anatto body paint (120). In Morobe Province, Papua New Guinea, the latex from the leaf stems is used to treat earache (288). In the East Indies the stem (with the bark removed) is used as a famine food (119). For a summary of the chemistry of this plant see Grenand *et al.* (1987).

CARYOCARACEAE

Caryocar villosum (Aublet) Pers.
[no voucher taken]
Tree in *terra firme* forest.
French Guiana and Amazonian Brazil (221).
warky (Waimiri Atroari). The fruits are boiled and the oily mesocarp (1.6% protein, 25.6% fat, 30.4% carbohydrate (4)) is eaten, often rubbed onto a piece of manioc 'bread'. The seeds are dried slowly over a fire and are also eaten. Slabs of the bark may be dried over a fire, burned, and salt extracted from the ashes. These trees are highly valued and may be left standing in swidden sites.
piquiá (local vernacular).

butter nut (English). The seeds are a source of edible oil (70%), which has also been employed in the manufacture of facial creams (201) and soaps (193). The Wayãpi Indians of French Guiana add this oil to their anatto body paint (120). In the vicinity of Manaus an infusion of the bark is used as a febrifuge and diuretic (261). It is also incorporated in a medicine used by the *caboclos* of the lower Amazon to treat fungal infections of the feet (10). This species, due to its relatively high vitamin A content, is said to prevent and cure eye disorders if eaten (193). Its occurrence is regarded by the Ka'apor Indians of Brazil as an indicator of soil fertility (29). The wood is used in heavy construction and boat-building (221), and in French Guiana is used for dug-out canoes (208). The fruits are eaten by saki monkeys (18).

CELASTRACEAE

Goupia glabra Aublet *2
[594]
Tree in *terra firme* forest.
warmia (Waimiri Atroari). The wood is much used for fuel, particularly for cooking meat. The fruits are eaten by toucans, parrots, macaws and monkeys.
cupiúba (local vernacular; = termite tree in Tupi). The Wayãpi Indians of French Guiana eat the fruits of this species (120). The Créoles of French Guiana use a decoction of the bark to cure toothache (122), and the use of a decoction of the leaves as a cure for eye infections has also been reported from French Guiana (128). In Guyana the juice of the inner bark is also used for toothache (97). In Colombia a decoction of the leaves is used as a hair dye (237) and as a treatment for cataracts (257). The timber is used for construction, charcoal etc. throughout Brazilian Amazonia (166), and is exported. In the Guianas this species is used for making canoes (239, 208). The fruits are eaten by spider monkeys (228), by saki monkeys (18) and by howler monkeys (120).

CHRYSOBALANACEAE

Couepia elata Ducke *2
[63]
Tree in *terra firme* forest.
Lower Rio Negro region, Brazil (216).
xuwi iyhy (Waimiri Atroari; *xuwi* = tinamou). The wood is used for arrow-heads. The fruits (?) are eaten by guans, tortoises and macaws.

Hirtella racemosa Lam. GR
[GR2706]
Small tree/treelet in *terra firme* forest.
Panama and Colombia through the Guianas and Amazonia to northern Bolivia (216).
pyrwa iepry (Waimiri Atroari; possibly a name but more probably a description of its use). The wood is used for arrow-heads.

The fruits are eaten by the Chácobo Indians of Bolivia (48). In French Guiana the trunks and branches are split into strips to make partition walls and wickerwork (208). The Taiwano Indians of Colombia use an oil extracted from the flowers to treat (fungal?) earache (257). The fruits are eaten by squirrel monkeys (271).

Licania apetala (E. Meyer) Fritsch *2
[146]
Tree in *terra firme* forest.
The Guianas and Amazonia (216).

xuwi iyhy (Waimiri Atroari; *xuwi* = tinamou). The wood is said to be extremely hard. The fruits (?) are eaten by guans, tortoises and macaws.

caraipé (local vernacular). The Puinave Indians of Colombia burn the bark, and add the silica-rich ashes to clay to strengthen pottery (254). The Tikuna Indians of Colombia apply poultices of the ground-up seeds to persistent skin sores (257).

Licania heteromorpha Benth. *3
[315]
Tree in *terra firme* forest.
Trinidad, Venezuela, the Guianas and Amazonian Brazil (216).

a'xa (Waimiri Atroari). Slabs of the (purple) inner bark are boiled in water, colouring it a deep red. This is mixed with tapioca to make *xipija*, which may be added to banana or 'buriti' (*Mauritia flexuosa*) drinks for *marba* ceremonies. The fruits are eaten by guans.

macucu fôfo, macucu do Rio Negro (local vernacular). The bark is used as a dye for ink and paint in Brazil (9), and the Wayãpi Indians of French Guiana use this to decorate their arrows (120). The Ka'apor Indians of Brazil use this species to treat mouth ulcers, paint gourds with its sap, and value its wood as a cooking fuel on account of its particularly hot flame (23). The *caboclos* of the lower Amazon drink an infusion of the bark as a cure for excessive menstrual bleeding, or use it as a wash for the same purpose (10). The Makú Indians boil the bark and then chew it to relieve toothache (257). A type of varnish may be prepared from the mesocarp of the fruits (73).

Licania hirsuta Prance *1
[386]
Tree in *terra firme* forest.
Brazilian Amazonia (216).

a'xa (Waimiri Atroari). See *L. heteromorpha* for uses etc.

—***Licania longistyla*** (Hook. f.) Fritsch *1
[609] [GR2782]
Tree in *terra firme* forest.
Amazonia and the Guianas (216).

kamda, pamda (Waimiri Atroari; *fide* 179). The bark is burned, and the silica-rich ashes added to clay to strengthen pottery. The fruits (?) are eaten by guans and tortoises.

are' (Waimiri Atroari). The fruits are edible. The wood is used for arrow heads. The fruits (?) are eaten by agoutis, pacas, tortoises and tapirs.

caraipé, uchi de cutia (local vernacular; *uchi* = various Humiriaceae, *cutia* = agouti). The fruits are reported to be edible when ripe (9).

Licania micrantha Miq. *2
[384]
Tree in *terra firme* forest.
The Guianas, Venezuela and Amazonia (216).

kamda, xuwi iyhy (Waimiri Atroari; *xuwi* = tinamou). The bark is burned, and the silica-rich ashes added to clay to strengthen pottery. The wood is used for arrow heads. The fruits (?) are eaten by guans, tortoises and macaws.

caraipé, macucu farinha seca, pintadinha (local vernacular; *farinha seca* = dry manioc flour, *pintadinha* = little painted one). This species furnishes particularly good charcoal (73), and its wood may be used for house-frames and paving blocks (96). The Wayãpi Indians of French Guiana use the split wood to make their walls, and eat the seeds (120). The young leaves and bark are eaten by spider monkeys (238), and the fruits by cutias and peccaries (120).

—*Licania oblongifolia* Standl. *5
[405]
Tree in *terra firme* forest.
Amazonian Brazil (216).
are' (Waimiri Atroari). The fruits are edible. The wood is used for arrow heads. The fruits (?) are eaten by agoutis, pacas, tortoises and tapirs.
macucu chiador*, *pajurárana (local vernacular).

Licania pallida Spruce ex Sagot *2
[222]
Tree in *terra firme* forest.
Venezuela and Brazilian Amazonia (216).
xuwi iyhy (Waimiri Atroari; *xuwi* = tinamou). The fruits (?) are eaten by guans, tortoises and macaws.
macucu sangue (local vernacular).

—*Licania* sp. *1
[454]
Tree in *terra firme* forest.
tapyrma iepy (Waimiri Atroari; = black tree). The wood is used for arrow heads.

—*Parinari excelsa* Sabine *1
[662]
Tree in *terra firme* forest.
Amazonian Colombia and eastern Venezuela through the Guianas and Amazonian Brazil to east central Brazil (216).
are' (Waimiri Atroari). The fruits are edible. The wood is used for arrow heads. The fruits (?) are eaten by agoutis, pacas, tortoises and tapirs.
pajurázinho (local vernacular). The wood is used in construction, and the pulp oil for perfuming hair (216). The Panare Indians of Venezuela eat the fruits (49). The ash of the bark and the wood yield a tannin, and the shell and pulp of the fruits yield a dye which is used in West Africa (285).

Parinari montana Aublet *1
[421]
Tree in *terra firme* forest.
French Guiana and northern Brazilian Amazonia (216).
mepri (mepre) kytawyna*, *kaxuhy ipe (Waimiri Atroari; *mepri* = tapir, *kytawyna* = *Duckesia* (Humiriaceae)). The fruits are edible. The fruits (?) are eaten by tortoises, tapirs, agoutis and pacas.
pajurá da mata*, *pajurá de pedra (local vernacular). The Andoque Indians of Colombia use the rough endocarp as a rasp (162). The wood is used in construction, and the seeds possess (unspecified) therapeutic properties (73). The latter are eaten by the Wayãpi Indians of French Guiana (120).

—*Parinari parvifolia* Sandw. *1
[344]
Tree in *terra firme* forest.
Guyana and eastern Amazonia (216).
no Waimiri Atroari name given.

COMBRETACEAE

Buchenavia grandis Ducke *1
[500]
Tree in *terra firme* forest.
Brazilian Amazonia (94).
katxirmia (Waimiri Atroari; *xirmia* = a large water snake). The fruits are edible, and are eaten by spider and howler monkeys, and by macaws.
tanimbuca (local vernacular).

Buchenavia parvifolia Ducke *1
[541]
Tree in *terra firme* forest.
Northern Brazilian Amazonia and southern Venezuela (94).
ki'mia (Waimiri Atroari). The fruits are eaten by macaws.
tanimbuca amarela (local vernacular; *amarela* = yellow). A decoction of the leaves, mixed with those of a *Trichomanes* species, is used by the Waiwai Indians of Guyana to check internal bleeding (9).

Buchenavia cf. ***viridiflora*** Ducke *1
[278]
Tree in *terra firme* forest.
Northern Brazilian Amazonia (94).
ki'mia (Waimiri Atroari). A decoction of the bark is taken for diarrhoea. Classified as *pitxeeni kinji'* (i.e. having medicinal application for babies, the meaning of which was not clearly understood). The fruits are eaten by macaws.

CONVOLVULACEAE

Ipomoea batatas (L.) Lam.
[no voucher taken]
Vine cultivated in swidden clearings.
Cultivated widely throughout the tropics; native of the Neotropics.
mepe (Waimiri Atroari). The tubers are cooked and eaten. At least three varieties are distinguished: *mepe, merepyna* (yellow) and *ianabia* (red). Grenand (pers. comm.) collected another variety named *iadadua*.
batata doce (local vernacular).
sweet potato (English). In the Philippines this plant is used as an antidiabetic (285).

—*Maripa* sp. *1
[533]
Woody liana in *terra firme* forest.
kiriri (Waimiri Atroari; = liana). A fish poison is prepared from the bark.
kiriri (Waimiri Atroari). The bark is used to treat snakebite.

The fruits of some *Maripa* species are edible; the Wayãpi Indians of French Guiana eat that of *M. scandens* Aublet (120).

CUCURBITACEAE

Citrullus vulgaris Schrad.
[no voucher taken]
Vine cultivated in swidden clearings; a recent introduction.
Cultivated worldwide, native to Africa.

marasia, miritxia (Waimiri Atroari; from Portuguese). The ripe fruits are edible.
melancia (local vernacular).
water melon (English). In Africa the young fruits are eaten as a vegetable; the ripe seeds are also eaten and an edible oil may be extracted from them (137). In Mexico a decoction of the leaves is taken for malaria, and in Central America the seeds are attributed diuretic properties and (when fresh) their juice is said to act as a vermifuge (187). In China the seeds are used in soups (170).

Cucurbita sp.
[no voucher taken]
Vine cultivated in swidden clearings.
xirimu (Waimiri Atroari; from Portuguese jerimum). The fruits are edible.
abóbora (local vernacular).
pumpkin (English). Pumpkin seeds are used to prepare a vermifuge in Brazil (231). They contain cucurbitine, an amino acid with toxic properties (44).

DILLENIACEAE

Doliocarpus brevipedicellatus Garcke *1
[339]
Woody liana in *terra firme* forest.
Central America and tropical South America (146).
akwa (Waimiri Atroari; perhaps a corruption of the Portuguese name). The stem yields potable water. The fruits are eaten by guans and oropendolas.
cipó de água (local vernacular; = water vine). In the vicinity of Manaus the root is attributed diuretic and purgative properties (261).

Pinzona coriacea Mart. & Zucc. *1
[89]
Woody liana in *terra firme* forest.
Central America, West India and northern South America (146).
akwa (Waimiri Atroari). The stem yields potable water. The fruits are eaten by guans and oropendolas.
cipó de água (local vernacular).

Tetracera amazonica Kubitzki
Woody liana in *terra firme* forest.
Brazilian Amazonia, vicinity of Manaus (155).
txika bixi kiriri (Waimiri Atroari; = sharp-bark vine). A decoction of the bark is drunk as a cure for snakebite.

Preliminary analysis of bark samples at the Royal Botanic Gardens, Kew (Jodrell Laboratory), have indicated the presence of alkaloids in this species (R. Nash, pers. comm.).

DIOSCOREACEAE

Dioscorea trifida L.
[no voucher taken]
Vine cultivated in swidden clearings.
Native to and widely cultivated throughout northern South America.
nepy, syhyna pyna (Waimiri Atroari). The tubers are cooked and eaten.
cará (local vernacular).

yam (English). This species, which does *not* contain diosgenin, may be used to treat asthma (66).

ELAEOCARPACEAE

The three species of *Sloanea* collected during this survey were given the same name (*kysiwyra*), and the same properties, i.e:

The plank-like buttress roots may be used to make canoe paddles. If a man steps on the fruits it was said that he will not find tortoises in the forest (see p. 42). The fruits are eaten by macaws and parrots.

Note: The local vernacular name applied to many species of *Sloanea* is 'urucurana' (= false *Bixa orellana*), on account of their spiny fruits. Interestingly, the Waimiri Atroari name also appears to be derived from that of *B. orellana.*

Sloanea guianensis (Aublet) Benth. vel aff. *1
[308]
Tree in *terra firme* forest.
Trinidad through Venezuela and Amazonia to Rio de Janeiro (88).
kysiwyra (Waimiri Atroari). See above for uses etc.

S. guianensis, which is known in Surinam as 'koesoeweran' (88), is used in eastern Peru for firewood and for paddles (88). The Galibi Indians of French Guiana are said to use the inner bark to wrap tobacco (208). The fruits are eaten by howler monkeys, squirrel monkeys and capuchins (271).

Sloanea laurifolia (Benth.) Benth. vel aff. *1
[575]
Tree in *terra firme* forest.
Amazonia, the Guianas, Venezuela and Trinidad (88).
kysiwyra (Waimiri Atroari). See above for uses etc.

Sloanea sp. *2
[538]
Tree in *terra firme* forest.
kysiwyra (Waimiri Atroari). See above for uses etc.

EUPHORBIACEAE

—*Alchorneopsis floribunda* (Benth.) Muell. Arg. *1
[361]
Tree in *terra firme* forest.
Northern Amazonia and Guyana (92).
kysekery (Waimiri Atroari). The wood is used for fuel. The fruits are eaten by tapirs.
warmaia (Waimiri Atroari). This species is used in the *karawyri kany* healing brew used at *marba* ceremonies (see p. 43). The fruits are eaten by parrots.

Croton lanjouwensis Jabl. *5
[428]
Tree in *terra firme* forest.
kysepe, kysepi (Waimiri Atroari). The inner bark (which smells of camphor or ginger) is rubbed on the skin to ease localized pain, and on the forehead for fever. It was said that if a person keeps a section of the inner bark clamped between the teeth for a

whole day, they will never suffer toothache again. The bark is a constituent of the *karawyri kany* healing brew used at *marba* ceremonies (see p. 43). The fruits are eaten by tortoises, tapirs, toucans, guans, macaws and parrots.

dima (local vernacular). Preliminary analysis of bark samples at the Royal Botanic Gardens, Kew (Jodrell Laboratory) has indicated the presence of alkaloids in this species.

—*Croton* sp. *1
[296]
Tree in *terra firme* forest.

kysiwyra (Waimiri Atroari). The fruits are eaten by macaws.

Note: The name *kysiwyra* was also applied to *Sloanea* spp. (Elaeocarpaceae) by the Waimiri Atroari. Although this may have been a misidentification, it is interesting to note that in the local vernacular, certain species of *Croton* are similarly known by the name which is commonly allocated to *Sloanea* (i.e. 'urucurana'). This alliance (or maybe confusion), which is presumably a result of similarity in the leaf venations, is even reflected in the scientific nomenclature, e.g. *Croton urucurana* Baill.

Hevea guianensis Aublet *1
[375]
Tree in *terra firme* forest.
Widespread in Amazonia (85).

kynyry ba (Waimiri Atroari). A dab of the latex from the bark, capped with a piece of leaf, is used to draw out parasitic botfly larvae from the skin. The fruits are eaten by macaws.

seringa (local vernacular). The Waorani Indians of Ecuador drink the latex as a fortifier, and use it to remove botfly larvae (81). This species produces low quality rubber (85), and in French Guiana its seeds are said to be good to eat (208).

—*Hyeronima* (?) sp. *1
[616]
Tree in *terra firme* forest.

no Waimiri Atroari name given. The wood is good for fuel.

parykaha (Waimiri Atroari). An infusion of the bark is attributed medicinal properties.

The inner bark of this species, another specimen of which was collected by W.M. on the Ilha de Maracá (Roraima State), emits a strong smell of wintergreen-type embrocation. The fruits of *H. laxiflora* M. Arg. are eaten by pigeons and toucans (120).

Manihot esculenta Crantz
[no voucher taken]
Shrub cultivated in swidden sites.
Widely cultivated throughout the tropics; native of the New World.

minja (Waimiri Atroari). This species forms the main staple crop of the Waimiri Atroari. At least 12 cultivars are recognized: *xuia pyny* ('bright' (yellow) flesh, best for making 'farinha' flour); *xuia iramemi*; *tiykadjidjika* (leaf entire); *domasi*; *sehe iala* (tall); *wasai*; *manjikwera* (pink petiole); *kybyma* (low); *maina* (red petiole); *iabwa*; *makaxira* and *topanoke*. The latter two are varieties of 'sweet' manioc (or 'macaxeira' in Portuguese), which can be boiled and eaten. The other ('bitter') varieties must be grated, washed and squeezed to remove the poisonous cyanogenic glycosides. The washed flour is generally cooked as large flat cakes (*mei*), but is sometimes roasted as 'farinha' flour (*sanja*). The suspended starch granules (*minja kaha*) are left to settle in the extracted fluid, removed, and dried in balls over the fire. Boiled in water, this starch produces tapioca which may be flavoured with other plants and eaten, or incorporated in certain beverages. The extracted fluid may be used as a fish poison.

mandioca brava, macaxeira (local vernacular).

manioc, cassava (English). In Brazil the poisonous extract ('tucupi') of the roots is boiled to render it harmless, and used as the basis of traditional fish and meat dishes. The Makuna Indians use this extract as a wash to treat scabies, and the Tikuna drink small quantities of it to cure diarrhoea (257). If left to putrefy, a type of maggot develops in the liquid which can be dried and used as a powerful poison (124). The leaves, which are eaten by Yurumanguy Indians (87), are used by the Wayãpi of French Guiana to make haemostatic plasters for arrow wounds (122), and similar plasters may be used to extract 'negro-worms' from the skin (124). Latex from the stem may be used as eye-drops to treat conjunctivitis (87).

GNETACEAE

—*Gnetum nodiflorum* Brongn. *1

[635]

Woody liana in *terra firme* forest.

Widespread throughout northern Amazonia (61).

kiriri (Waimiri Atroari; = liana). No uses given.

itúa (local vernacular). In Venezuela the resin is burned as incense (9), and the fruits are roasted and eaten in the vicinity of Iquitos, Peru (279). The Puinave Indians of Colombia apply a hot decoction of the bark (externally) to muscle and tendon injuries, and the Tiriyó of Surinam use a decoction of the crushed leaves as a wash to treat headaches (257). A strong rope can be made from the fibrous bark (62).

GRAMINEAE

Coix lacryma-jobi L.

[No voucher taken]

Large grass presumably grown in swidden sites.

South East Asia, cultivated widely in the tropics.

Waimiri Atroari name not recorded. The seeds are sometimes included in bead necklaces and bracelets.

lágrima de nossa senhora (local vernacular).

Job's tears (English). The seeds are used as a (rather poor) food source in some Asian countries, and the plant has been attributed a variety of medicinal properties including vermifuge and gonorrhea cure (decoction of the roots) and catarrh cure (decoction of the glume) (285). In Colombia a decoction of the leaves is used externally to cure rheumatism, and internally as an antispasmodic and diuretic (198).

Gynerium sagittatum (Aublet) Beauv.

[793]

Giant grass cultivated in dense stands in swidden sites.

Mexico and the West Indies to Peru and Brazil (67).

mapyry (Waimiri Atroari). The long stems of the inflorescenses are straightened over a fire and used as arrow shafts (see p. 27). Split lengths of the same are used to make the *katba* calendars sent out as invitations to *marba* ceremonies.

flecheira (local vernacular; = arrow-plant, arrow-provider). A sweet juice can be squeezed from the stem of this species (208). The root possesses diuretic properties (198). The Island Caribs of Dominica use the young stem for basketry, and the leaf midrib for making hats. The 'heart' (young unopened leaves) is used as to prepare a shampoo which promotes hair growth, and the leaves are used for thatch (136). They are also sometimes used as emergency fodder for livestock (54). In rural Peru the stems are used to construct the ceilings of houses (192). The pith of the stem is eaten by capuchin monkeys (271).

Saccharum officinarum L.
[no voucher taken]
Herb cultivated in swidden sites.
Cultivated throughout the tropics; first domesticated in New Guinea.
awaxe, awaxi (Waimiri Atroari). The stems are chewed for their sugar content. At least five varieties are recognized: *karimyke*; *tykurykwake*; *awaxe*; *tyrawyke* and *warybake*. One informant stated that the 'black' cane can cause 'madness'.
cana de acúcar (local vernacular).
sugar cane (English). The root is a diuretic, and there are records of this species being used as a fish poison (87). In Samoa the leaf ash is used to treat sore eyes (276). A wax may be obtained by boiling up the stems (137). Bagasse, the fibrous residue left after sugar extraction, may be used to manufacture board, plastics, charcoal, paper and several other products (259).

GUTTIFERAE

Clusia grandiflora Splitg. GR
[GR2786]
Epiphyte on island in igapó forest.
kypy' (Waimiri Atroari). The latex (*iynyry*) is used to treat fresh wounds.

In French Guiana the Palikur Indians use the latex as a plaster, and drink a decoction of the aerial roots to treat 'blesses' (a type of internal abcess), and the Créoles use a decoction of the plant as an external wash (122). The fruit are eaten by large bats (*Vampyrum spectrum*) (120). In Surinam the resin from the fruits is applied to fresh wounds (95).

Clusia insignis Mart.
[686]
Hemi-epiphyte/strangler in *terra firme* forest.
kypy' (Waimiri Atroari). Latex from the stem is put into the eyes to 'increase the likelihood of finding tortoises' (see p. 42).
apuí (local vernacular). The fruits is used in the vicinity of Manaus as an emetic and diuretic (261). Natives of the upper Río Negro of Colombia use the latex to treat toothache (253). The Kayapó Indians of Brazil plant this species as a source of medicine (13).

Clusia sp.
[801]
Hemi-epiphyte/strangler in *terra firme* forest.
marabia (Waimiri Atroari). The woody bases of fallen male flowers are dried and burned. Inhalation of the smoke relieves pain and fever. Traditionally these were sometimes included in the *kwyie* bead aprons worn by the women.

Symphonia globulifera L.f. *1
[558]
Tree in *terra firme* forest.
Tropical South and Central America and the West Indies (95).
manji (Waimiri Atroari). Latex from the bark is mixed with the boiled latex of *Couma macrocarpa* (Apocynaceae) or *Brosimum parinarioides* (Moraceae), and set into blocks. The blocks of resin are used in arrow fabrication (see p. 29) and as a general adhesive. Fresh latex is painted directly onto the skin for decoration, where it leaves a brown mark which lasts for several days. The fruits are eaten by parrots.

anani (local vernacular). In Brazil the oil from the seeds is used to treat skin diseases (277). The Makuna Indians of Colombia burn the bark and apply the ashes to wounds (254), and Créoles of French Guiana use the latex to seal them (122). In the Guianas an infusion of the bark is used for bathing ulcers (97). The Indians of the Río Apaporis (Colombia) use the ashes to treat ulcers of the abdomen and legs (257). The *caboclos* of the lower Amazon ingest the latex as a fortifier, or as a purgative for pregnant women (10). The latex of this species is taken orally as a contraceptive by the Ka'apor of Brazil (28), who also burn the resin to light their houses (23). The wood is used for construction, furniture and paper pulp in Brazil (166). The latex yields a khaki dye which is used on leather in Guyana (97). The flowers are eaten by spider monkeys (238), and the fruits and/or flowers by golden parakeets (28).

HELICONIACEAE

Heliconia sp.
[no voucher taken]
Giant herb in regions of poor drainage alongside road.
paryry (Waimiri Atroari). The long petioles are crushed to soften them, and used as temporary carrying straps or cordage.

HIPPOCRATEACEAE

—*Cheiloclinum hippocrateoides* (Peyr.) A.C. Smith **1*
[724]
Woody liana in *terra firme* forest.
Guiana to Amazonian Peru, Bolivia and Brazil (264).
kiriri (Waimiri Atroari; = liana). No uses given.

Salacia cf. ***impressifolia*** (Miers) A.C. Smith **1*
[109] [GR2704]
Woody liana in *terra firme* forest.
Amazonian Brazil and adjacent Peru and Bolivia (264).
arawyta xiky ba (Waimiri Atroari; *arawyta* = howler monkey, *xiky ba* = bladder). The fruits are edible, and are eaten by monkeys.
gogó de guariba (local vernacular; = howler monkey's Adam's apple).

Salacia cf. ***elliptica*** G. Don (agg.) **1*
[333]
Woody liana in *terra firme* forest.
Eastern Brazil, Paraguay, Guyana and Colombia (264).
arawyta xiky ba (Waimiri Atroari). See species above.

Tontelea cf. ***coriacea*** A.C. Smith **1*
[287]
Woody liana in *terra firme* forest.
Guyana, Amazonian Peru (264) and Brazil (?).
arawyta xiky ba (Waimiri Atroari). See species above.
Note: Grenand (pers. comm.) has questioned the palatability of this species, and since it was collected in a sterile state it may perhaps have been confused with another (morphologically similar) species of Hippocrateaceae.

HUMIRIACEAE

Duckesia verrucosa (Ducke) Cuatrec. *1
[185]
Tree in *terra firme* forest.
Central Amazonia, in the vicinity of Manaus (77).
kytawyna (Waimiri Atroati, *kyta* = low). The fruits are edible (boiled), and are eaten by tapirs, collared peccaries and agoutis.
uxí coroa, uxícuruá (local vernacular).

Sacoglottis guianensis Benth. *2
[179]
Tree in *terra firme* forest.
Amazonian Brazil, Colombia and Venezuela, and the Guianas (77).
kymdyry (Waimiri Atroari). The fruits are eaten by macaws, pacas, peccaries, tortoises, tapirs and agoutis.
macucu murici, achuá (local vernacular). An extract of the bark of this species is used as a tanning agent for leather, and as a dye for calabash bowls in the Brazilan Amazon (231, 120). The wood is used for construction (230).

Sacoglottis matogrossensis Malme var. ***subintegra*** (Ducke) Cuatrec. *1
[716]
Tree in *terra firme* forest.
Amazonian Brazil and Colombia (77).
kymdyry (Waimiri Atroari). An infusion of the inner bark is drunk as a cure for coughs, and the bark may be chewed for the same purpose. The fruits are eaten by macaws, pacas, peccaries, tortoises, tapirs and agoutis.
achuá, uxí preto (local vernacular; *preto* = black). The fruits are edible (261).

ICACINACEAE

—***Emmotum*** cf. ***fagifolium*** Desv. ex Hamilt. *1
[266]
Tree in *terra firme* forest.
markaka (Waimiri Atroari). It was said that the wood is used by *kamia* (white man) for construction, but that the trees were too large to be of use to the Waimiri Atroari.

Humirianthera rupestris Ducke
[783]
Woody vine in *terra firme* forest and secondary vegetation.
sywa', sawa (Waimiri Atroari). The bulky subterranean tuber was an important food source in the past, and is now recognized as a 'famine food', available should the manioc crop fail. The tuber must be grated and rinsed in much the same way as manioc, to remove the poisonous compounds contained therein. A fish poison is prepared from the stem of this species. According to Grenand (pers. comm.) it is left to grow in the swidden clearings and can thus be regarded as semi-cultivated.
batata mairá (local vernacular; *batata* = potato). Humirianthenolides (degraded diterpenoids) have been found in the tubers (297). According to the Wayãpi Indians of French Guiana, planting the tubers of *H. ampla* (Miers) Baehni in swidden sites favours the growth of manioc (120).

LAURACEAE

Note: Almost all of the Lauraceae collected during this study were identified as either *penji* or *awa'ba*; names which may perhaps be interchangeable (see p. 18). They were all recognized as providing suitable timbers for construction. The particular uses with which a species was attributed may have been influenced by the size and shape of the individuals examined. Their fruits were said to be eaten by macaws, parrots, guans, trumpeters and toucans. The local vernacular name generally applied to the Lauraceae is 'louro'.

Lauraceous tree species are not easy to distinguish from each other in the field; Grenand (pers. comm.) noted a lot of confusion over this family among the Wayãpi and Palikur Indians of French Guiana. This should be borne in mind when considering the following data.

Aniba williamsii O. Schmidt *1
[91]
Tree in *terra firme* forest.
Surinam, Peru and Amazonian Brazil (156).
penji (Waimiri Atroari). The trunks are used for house-posts.

In Loreto, Peru, this species is attributed minor commercial value (204).

Endlicheria verticillata Mez *1
[164]
Tree in *terra firme* forest.
penji*, *awa'ba (Waimiri Atroari). The trunks are used for house-posts and beams.

Licaria aurea (Hub.) Kosterm. *1
[545]
Tree in *terra firme* forest.
penji (Waimiri Atroari). The trunks are used for house-posts.
louro aritu amarelo (local vernacular).

Licaria canella (Meissn.) Kosterm. *1
[168]
Tree in *terra firme* forest.
penji (Waimiri Atroari). The trunks are used for house-posts.

In Guyana the wood is used for boat construction, dug-out canoes and furniture (96). The Wayãpi Indians of French Guiana also use the wood to make war-clubs (122). The Créoles (also of French Guiana) use a decoction of the bark as an aphrodisiac, and use the bark itself as a substitute for cinnamon (122).

Licaria guianensis Aublet *5
[143] [GR2719]
Tree in *terra firme* forest.
awa'ba*, *penji (Waimiri Atroari). The trunks of large individuals are used for dug-out canoes.
louro aritu (local vernacular).

Licaria sp. *1
[205]
Tree in *terra firme* forest.
awa'ba*, *penji (Waimiri Atroari). The trunks are used in house construction.

Mezilaurus sp. *1
[149]
Tree in *terra firme* forest.
tawyba (Waimiri Atroari; possibly a corruption of the local vernacular). The trunks are used for dug-out canoes, and for making paddles.
itaúba (local vernacular; = rock-tree in Tupi). *Mezilaurus* species yield the preferred (rot-resistant) timber for the construction of passenger boats on the Rio Negro in Brazil.

Ocotea amazonica (Meissn.) Mez vel aff. *2
[86]
Tree in *terra firme* forest.
Amazonia (236).
penji (Waimiri Atroari). The trunks are used to make dug-out canoes.

Ocotea canaliculata (Rich.) Mez *1
[719]
Tree in *terra firme* forest.
Northern Brazilian Amazonia and the Guianas (235).
awa'ba bahnja*, *penji (Waimiri Atroari; *bahnja* = small). The trunks are used for house-posts.

In Guyana the wood is used for internal construction and light furniture (96).

Ocotea rubrinervis Mez vel aff. *1
[207]
Tree in *terra firme* forest.
Venezuela, Panama and Peru (236), and Brazil (?).
penji (Waimiri Atroari). The trunks are used in house construction.

Ocotea sp. *5
[666]
Tree in *terra firme* forest.
awa'ba*, *penji (Waimiri Atroari). The trunks are used for house construction. Large specimens may be used to make dug-out canoes.

Rhodostemonodaphne cf. ***grandis*** (Mez) Rohwer *2
[406]
Tree in *terra firme* forest.
Amazonia, Venezuela and the Guianas (236).
awa'ba tapyrma*, *penji (Waimiri Atroari; *tapyrma* = black). The trunks of large individuals are used for dug-out canoes, and those of small ones for house-posts. The inner bark is highly aromatic.

Genus undetermined *1
[726]
Tree in *terra firme* forest.
penji (Waimiri Atroari). The trunks are used in house construction.

LECYTHIDACEAE

Note: The Lecythidaceae occur in considerable abundance and diversity in the forest studied. Certain members of this family, such as *Bertholletia excelsa*, *Eschweilera rhododendrifolia*, *Couratari stellata* and *Lecythis pisonis*, are very distinctive in the field and were named with ease and certainty by the Waimiri Atroari informants. Most of the *Eschweilera* species, many of which are morphologically very similar, were lumped into the *doma* category. There was, however, a certain amount of confusion over the nomenclature and properties of some of the other Lecythidaceae.

Bertholletia excelsa Humb. & Bonpl.
[no voucher taken]
Tree in *terra firme* forest.
Amazonian Colombia, Venezuela, Peru, Bolivia and Brazil, and the Guianas (185).
tetky (Waimiri Atroari; a Carib name). The seeds are eaten. Fibres from the inner bark are sewn into loops and used for carrying babies. This species is highly valued, and is generally left standing in swidden sites. In the past the Crichaná Indians (see p. 13) used the woody capsules (pyxidia) as food containers, and prepared an evil-smelling 'butter' from rotten 'castanhas' pounded with fish (34).
castanha do Pára (local vernacular; = Pará chestnut).
Brazil nut (English). The Wayãpi Indians of French Guiana believe that this species was given to them by the agouti (120). The wood is of high quality, but rarely used because of the value of the tree as a fruit-bearer (166). Smoke from the burning capsules is reputedly good for coagulating rubber (185). Oil from the seeds may be used for cooking (219), and for making soap (163). The *caboclos* of the lower Amazon use an infusion of the leaves to wash wounds, and drink a mixture of the fruit husk with water to cure anaemia (10). In Manaus (Amazonas, Brazil) the bark is sold in the market as a medicine for liver complaints (WM pers. obs.), and an infusion of the leaves is taken for colic and inflamed ovaries in the same region (40). The Yanomami Indians used to build canoes from the bark of this species, which was removed entire from the trunk (69). The Boni negroes of the Maroni River, Cayenne, use the sap-wood (bruised and impregnated with resin) for caulking canoes (239). The capsules, ground to a fine powder, may be used as the basis for 'Bakelite'-type mouldings or for linoleum (137). The seeds are eaten by agoutis (220).

Couratari longipedicellata W. Rodrigues *1
[209]
Tree in *terra firme* forest.
Central Amazonian Brazil (185).
tyrwe (Waimiri Atroari). The fruits are eaten by macaws.

—*Couratari multiflora* (Smith) Eyma vel aff. *1
[140]
Tree in *terra firme* forest.
Venezuela, the Guianas and Amazonian Brazil (185).
dwoma, doma (Waimiri Atroari). Strips of the fibrous innner bark are used as carrying straps. The fruits are eaten by macaws.
kamtetky (Waimiri Atroari).
tauari branco, tauari amarelo (local vernacular). In French Guiana the Palikur and Wayãpi Indians use the inner bark to wrap cigars, to which they attribute narcotic properties (122).

Couratari stellata A.C. Smith *2
[550]
Tree in *terra firme* forest.
The Guianas, western and central Amazonian Brazil, and Amazonian Colombia and Venezuela (185).
anasa, iryry (Waimiri Atroari). Strips of the fibrous inner bark are used as carrying straps. The fruits are eaten by macaws, deer and *meiyra* (unidentified animal).

tauari (local vernacular). The inner bark may be used as cigarette paper (185), and the wood possesses particularly good elastic properties (233). The fruits are eaten by spider monkeys (238) and saki monkeys (181).

—*Eschweilera amazonica* Knuth *1
[364]
Tree in *terra firme* forest.
Central and eastern Brazilian Amazonia (185).
anasa (Waimiri Atroari). The fruits are eaten by macaws.
tyrwe (Waimiri Atroari).
matámatá preto, matámatáci (local vernacular). The inner bark furnishes useful fibres (2).

Eschweilera bracteosa (Poeppig ex Berg) Miers *3
[736]
Tree in *terra firme* forest.
Widely distributed throughout central and western Amazonia (185).
dwoma, doma (Waimiri Atroari). Strips of the fibrous inner bark are used as carrying straps. The wood is used in house construction. The fruits are eaten by deer, agoutis, macaws and *meiyra* (unidentified animal).
matámatá (local vernacular).

Eschweilera coriacea (A. DC.) S. Mori *53
[762] [GR2716]
Tree common in *terra firme* forest.
Eastern Panama, northwestern Colombia, and throughout the Guianas and the Amazon basin (185).
dwoma, doma (Waimiri Atroari). See *E. bracteosa* for uses etc.
matámatá, matámatá branco (local vernacular). The Ka'apor Indians of Brazil tie the inner bark around wounds as a bandage to hasten healing (23), and the Asurini use it to treat skin infections (26). The Tembé also use this species for wounds, and maintain that its wood burns slowly and is thus particularly suitable for roasting manioc-flour (24).

Eschweilera grandiflora (Aublet) Smith *21
[735] [GR2744]
Tree in *terra firme* forest.
French Guiana, eastern and western Amazonian Brazil, and Amazonian Peru (185).
dwoma, doma (Waimiri Atroari; ***anasa*** fide Grenand). See *E. bracteosa* for uses etc.
matámatá, castanha vermelha, tibira (local vernacular). In French Guiana the seeds are said to be good to eat (208).

Eschweilera rhododendrifolia (Knuth) A.C. Smith *10
[251]
Tree in *terra firme* forest.
Central Amazonia around Manaus, Brazil (185).
xirby, kyxirby (Waimiri Atroari). The trunks are used for houseposts. The wood is good for cooking manioc. The fruits are eaten by macaws, pacas and agoutis.

—*Eschweilera tessmannii* Knuth *1
[740]
Tree in *terra firme* forest.
Central and Western Amazonia (185).

xirby*, *kyxirby (Waimiri Atroari). See *E. rhododendrifolia* for uses etc.
ripeiro*, *ripeiro vermelho (local vernacular).

Eschweilera sp. aff. ***tessmannii*** Knuth *1
[625]
Tree in *terra firme* forest.
dwoma*, *doma (Waimiri Atroari). See *E. bracteosa* for uses etc.

—***Eschweilera*** sp. nov. *3
[204]
Tree in *terra firme* forest.
anasa (Waimiri Atroari). Strips of the fibrous inner bark are used as carrying straps. The fruits are eaten by macaws.
tyrwe (Waimiri Atroari).

Eschweilera sp(p). *5
[151, 211, 309, 480, 584]
Tree in *terra firme* forest.
dwoma*, *doma (Waimiri Atroari). See *E. bracteosa* for uses etc.

—***Gustavia elliptica*** S. Mori *3
[162]
Tree in *terra firme* forest.
Central Amazonia, common around Manaus, Brazil (219).
no Waimiri Atroari name given.
castanha fedorenta*, *castanharana (local vernacular; = foetid chestnut, false chestnut).

Lecythis pisonis Cambess. *1
[231]
Tree in *terra firme* forest.
Coastal Brazil and Amazonia (185).
awanma (Waimiri Atroari). The large seeds are edible. A fish poison is prepared from the bark (although it is said not to be as powerful as some other fish poisons used). In the past the woody capsules (pyxidia) were used as food containers (34). The fruits are eaten by tapirs, macaws and capuchin monkeys.
castanha sapucaia (local vernacular).
paradise nut (English). An infusion of the leaves or of the bark is used by the *caboclos* of the lower Amazon to treat itches (10). The bark contains tannins and saponins (but no alkaloids), and may be used to treat coughs (89). The wood is used for construction (230). The Wajakoele Indians of Surinam use the bast fibre for their loin cloths (239).

Lecythis retusa Spruce ex Berg *1
[262]
Tree in *terra firme* forest.
Central Brazilian Amazonia (185).
kamtetky (Waimiri Atroari). This species was said to be the source of a fish poison. The fruits are eaten by macaws and cutiaras.
castanha jaraná*, *jaraná (local vernacular).

—***Lecythis*** sp. *1
[197]
Tree in *terra firme* forest.
kamtetky (Waimiri Atroari). See *L. retusa* for uses etc.

LEGUMINOSAE

Acacia altiscandens Ducke   *1
[373]
Woody liana in *terra firme* forest.
kiriri kararapy (Waimiri Atroari). The fruits are eaten by parrots.

Andira sp.
[790]
Tree at margin of damp 'baixada' depression in forest.
ka'wa (Waimiri Atroari). Scrapings of the inner bark are mixed with a little water or saliva, and squeezed onto the skin to relieve localized aches and pains. According to one source this species may also be used as a vermifuge.

Several species of *Andira* are used as vermifuges in Brazil, including *A. anthelminthica* Benth. which is also known as 'mata barata' and is good for poisoning cockroaches (J.A. Ratter, pers. comm.). The name *Andira* comes from the Tupi name for bat, presumably because bats eat the fruits of most members of this genus.

—*Andira* sp. vel gen. aff.   *1
[298]
Tree in *terra firme* forest.
kymara (Waimiri Atroari). The wood is used in house construction. The bark is reputed to be poisonous.

In the Guianas, certain species of *Andira* are used as fish poisons (97).

—*Andira* sp. vel gen. aff.   *2
[539]
Tree in *terra firme* forest.
no Waimiri Atroari name given. The trunks are used for house-posts.
tykepra iepy (Waimiri Atroari). The fruits are edible, and are eaten by trumpeters and capuchin monkeys.

Bauhinia guianensis Aublet   *3   (= *Bauhinia splendens* Kunth)
[392]
Woody liana in *terra firme* forest.
The Guianas, Venezuela, Colombia and Guadeloupe (11), and northern Amazonian Brazil.
wiamy dydwe, (Waimiri Atroari; *wiamy* = tortoise). An infusion of the bark is taken to cure excessive menstrual bleeding and diarrhoea. Mawa, the first Indian, used this vine to climb into the sky (see p. 42).
escada de jabotí (local vernacular; = tortoise's ladder). An infusion of the root bark is drunk in the vicinity of Alter do Chão (Pará, Brazil) to cure diarrhoea and amoebic dysentery (50). A decoction of the stem is used on the Putumayo River, Colombia, as a febrifuge and as a treatment for leprosy (107). An infusion of the bark, or the sap itself, is taken by the Tiriyó Indians of Brazil to cure coughs (63). The stem is wrapped around the abdomen by the Kayapó Indians of the Xingu (Brazil) to cure excessive menstrual bleeding (90). The Tukuna (Tikuna) Indians of Colombia drink an infusion of the stem to treat kidney ailments (111), and the *caboclos* of the lower Amazon take an infusion of the seeds for stomach ache and dysentery (10). The Taiwano Indians of Colombia attribute diuretic properties to these seeds (257). This species is used as a fish poison by the Caribs of Guyana (9).

Bocoa viridiflora (Ducke) Cowan *3
[269] [GR2732]
Tree in *terra firme* forest.
tyny' (Waimiri Atroari). The trunks are used for house-posts.
muirajiboia preta (local vernacular; = black boa constrictor wood). The wood of this species is extremely hard. The Wayãpi Indians of French Guiana use a very similar name (tunu'y), which means 'stick-tree', for the closely related legume *Candolleodendrum brachystachyum* (DC.) Cowan (120).

Clathrotropis macrocarpa Ducke *72
[542] [GR2726]
Tree in *terra firme* forest.
kaba (Waimiri Atroari). The large seeds provide a starchy food, currently regarded as a famine food but apparently an important dietary component in the past. They must first be grated and washed to remove the poisons which they contain. The wood is good for fuel. The fruits are eaten by agoutis.
cabari (local vernacular). The bark has been reported to be the source of a fish poison (9). The seeds contain toxic quinolizidine alkaloids (G.P. Lewis, pers. comm.). In Colombia the leaves are used to wrap up manioc flour, to keep away weevils (1).

Derris amazonica Killip
[782]
Woody liana in *terra firme* forest.
Surinam and northern Brazil (Pará and Amazonas) (11).
waiwai karda (Waimiri Atroari; *waiwai* = Waiwai Indians, *karda* = fish poison). The stem of this species is used to prepare a fish poison.
timbórana, timbó branco (local vernacular). The Chácobo Indians of Bolivia use this species as a fish poison, but claim that its effect is relatively weak (48). The roots contain 0.3% rotenone (153).

Dipteryx odorata (Aublet) Willd. *1
[761]
Tree in *terra firme* forest.
The Guianas and northern Amazonian Brazil (11).
kyma, kuma (Waimiri Atroari). The fruit (or seed?) is clamped between the teeth to relieve toothache.
cumaru (local vernacular). The seeds, which are eaten by the Wayãpi Indians of French Guiana (120), contain 16% protein and 56.4% oil (201). The oil is used to treat dysentery in French Guiana (128), and as a cure for a multitude of ailments by the inhabitants of Alter do Chão, Pará, Brazil (50). The *caboclos* of the lower Amazon use the oil to cure sore throats (10), and the Tembé Indians use it for pneumonia (24). An infusion can be taken to treat cardiac and respiratory problems, including asthma (39). The oil contains coumarin, which has been employed as a vanilla substitute, a whisky flavouring, a tobacco aromatizer and a perfume ingredient (201). In Guyana a decoction of the bark is used as a febrifuge (97), and its resinous gum is used to treat sore throats and in tanning. The Palikur Indians of French Guiana employ this species in a fortifying bath for babies (122). The bark contains lupeol and betulin (189), and the seed coat contains diterpenoids (112). A coumarin derivative, dicoumarol, has been used as an artificial anticoagulant and as a rodenticide (186). The timber is extremely durable and heavy, and is used in construction and shipbuilding etc. in Brazil (166). The Wayãpi use the endocarp to make blunt arrow-heads for hunting small birds (120). The young leaves are eaten by spider monkeys (238).

Hymenaea sp.
[no voucher taken]
Tree in *terra firme* forest.
ximri (Waimiri Atroari). The sweet floury mesocarp is eaten. Resin exuded from the trunk may be used to coat (glaze) pottery.
jatobá (local vernacular; = hard-fruited tree in Tupi). The bark of certain *Hymenaea* species is used in the Guianas for making canoes (97).

Inga brevialata Ducke *1
[232]
Tree in *terra firme* forest.
arawyta pana, iukwiri (Waimiri Atroari; = howler monkey's ear). The fruits are edible (the sweet pulp around seeds). The fruits are eaten by parrots, macaws and howler monkeys.
ingá (local vernacular).

Inga capitata Desv. *1
[578]
Tree in *terra firme* forest.
Brazil, French Guiana and Surinam (152).
iukwiri (Waimiri Atroari; this name is also used for 'beans'). See *I. brevialata* for uses etc.
ingá costela (local vernacular; *costela* = rib). This species is called 'apoeroekoeni' by the Caribs of Surinam (152) — see the Waimiri Atroari name for *I.* cf. *aggregata*. The Wayãpi Indians of French Guiana use a decoction of the leaves to cure female infertility (122). The fruits are eaten by spider monkeys (238).

Inga edulis Mart.
[no voucher taken]
Tree cultivated around villages.
Widely cultivated throughout, and native to, South America.
akai (Waimiri Atroari), ***kynabra*** (*fide* Grenand). The fleshy pulp around the seeds is eaten.
ingá-cipó (local vernacular).
ice cream bean (English). The Wayãpi Indians of French Guiana eat the fruits of this species from wild stock, and maintain that its wood is good for fires (120). The fruits are eaten by capuchin and squirrel monkeys (271).

Inga cf. ***fagifolia*** (L.) Willd. *2
[226]
Tree in *terra firme* forest.
iukwiri (Waimiri Atroari). See *I. brevialata* for uses etc.

Inga paraensis Ducke *1
[512]
Tree in *terra firme* forest.
iukwiri (Waimiri Atroari). See *I. brevialata* for uses etc.
ingá branco (local vernacular; = white ingá). The Wayãpi Indians of French Guiana use the sap of this species as a mordant for a dye (120).

Inga rubiginosa (Rich.) DC. GR
[GR2756]
Tree in *terra firme* forest.
French Guiana, Demerara, Surinam and Brazil (152).
arawyta pana (Waimiri Atroari; = howler monkey's ear). The fruits (which are hairy) are edible, and are eaten by monkeys.

ingá (local vernacular). The Kubeo Indians of Colombia dry and powder the flowers and sniff them to clear nasal congestion (257). The Wayãpi Indians of French Guiana call this species 'howler monkey ingá', and also eat its fruits (120). In Surinam its fruits are also compared to spider monkeys (152).

Inga spectabilis (Vahl) Willd. GR
[GR2707]
Tree in *terra firme* forest.
Northern tropical South America through Panama to Costa Rica (54).
arawyta pana (Waimiri Atroari; *arawyta* = howler monkey). The fruits are edible (sweet pulp around seeds). The fruits are eaten by monkeys.

Inga cf. ***aggregata*** G. Don
[786]
Tree in *terra firme* forest.
Peru, Guyana and Surinam (152), and northern Brazil (?).
apirikini (Waimiri Atroari). Slabs of the bark are cut from the trunk, and the inner surface (cambium) with its sticky pink sap, is scraped off. This is mixed with soot from the burned resin of various *Protium* species (Burseraceae), to produce a black paint used to dye strips of *Ischnosiphon* (Marantaceae) stem for decorative basketry, and to decorate arrows.
ingá (local vernacular). *I. alba* (Sw.) Willm. and *I. pezizifera* Benth. are used for the same purpose by the Indians of French Guiana (122). In fact several closely related species are employed for this purpose in the Guianas, and are known by very similar names by most Carib tribes (Grenand, pers. comm.).

—*Machaerium quinata* (Aublet) Sandw. *1
[148]
Woody liana in *terra firme* forest.
Guyana and Surinam (11), and northern Amazonian Brazil.
kiriri (Waimiri Atroari; = liana). The copious red sap is used to treat lesions of the tongue. It was said by one informant that an arrow poison had been produced in the past from this sap (taken from young individuals), by boiling it until thick and sticky and applying it directly to the arrow-head. Baines (pers. comm.) suggests that the latter may be a result of Waiwai Indian influence, and Grenand (pers. comm.) suggests that the species was confused by the informant with a member of the Connaraceae, many of which also possess red sap and composite leaves and are poisonous.

The Wayãpi Indians of French Guiana prepare a dye from this species (120).

Macrolobium unijugum (Poeppig & Endl.) Cowan ssp. ***unijugum*** *2
[107] [GR2724]
Tree in *terra firme* forest. Northern Brazilian Amazonia, Peru and southern Venezuela (75).
parakwy (Waimiri Atroari). Strips of the fibrous inner bark are used for carrying game. The fruits are eaten by pacas.
faveira (local vernacular).
Note: the same name was said to refer also to a species which grows in profusion alongside the river, and which is probably used more often for fibre than is this species. At least two species of *Macrolobium* were seen in the igapó forests, including *M. acaciifolium* (Benth.) Benth. whose (presumably fibrous) bark is used by the Wayãpi Indians of French Guiana for canoes (120).

Marmaroxylon* (*Pithecellobium*) *racemosum (Ducke) L. Rico ined. *1
[571]
Tree in *terra firme* forest.
kiowyry (Waimiri Atroari). An infusion of the bark is drunk to cure coughs. The leaf is a constituent of the *karawyri kany* healing brew used at *marba* ceremonies (see p. 43). Spider monkeys and macaws sleep in the branches.
angelim rajado (local vernacular). The wood is highly valued for its beauty around Manaus (261), and may be used for parquet floors and furniture (96). This species provides food (fruits?) for tapirs (24).

Marmaroxylon* (*Pithecellobium*) *ramiflorum (Benth.) L. Rico *2
[571] [GR2749]
Tree in *terra firme* forest.
Brazil, Venezuela, Peru and Panama (229).
kiowyry (Waimiri Atroari). Spider monkeys and macaws sleep in the branches.

Parkia nitida Miq. (= *P. oppositifolia* Spruce ex Benth.)
[431]
Tree in *terra firme* forest.
Southern Panama through eastern Venezuela and the Guianas to Amazonia (139).
tawyxi xsypra (Waimiri Atroari). Pieces of the inner bark (which smell strongly of 'Vick' ointment) are rubbed on the forehead to relieve headache and fever.
fava benguê (local vernacular). An infusion of the bark is drunk by the Makú Indians of northern Brazil to cure dysentery (254), and can be taken for headaches and rheumatic pains (261). An infusion of the bark is astringent, antihaemorrhagic, and is used for washing cuts and ulcers (74). The *caboclos* of the lower Amazon bathe their babies in it (10). The Bora Indians of Peru and the Wayãpi Indians of French Guiana regard the fruits of this species as edible (102, 120). The mature seeds, analysed for their commercial potential, were found to contain proteins rich in triptophans (158). The wood is used for construction and paper pulp in Brazil (166). The flowers are eaten by spider monkeys (238), and the fruits by woolly monkeys and howler monkeys (144).

Parkia panurensis H.C. Hopkins *1
[445]
Tree in *terra firme* forest.
Widespread throughout Amazonia (139).
sawyraha (Waimiri Atroari). Spider monkeys sleep among the branches, and the fruits are eaten by macaws.
faveira, visgueiro (local vernacular). The resin around the seeds is edible (139), and the gummy resin exuded from the bark is reputed to possess medicinal properties (73).

Parkia velutina Benoist *1
[367] [GR2778]
Tree in *terra firme* forest.
Amazonian Brazil and Peru, French Guiana, eastern Venezuela and Colombia west of the Andes (139).
mara' (Waimiri Atroari). An infusion of the bark is used as *pitxeni kipytxiwy* (= baby medicine, exact meaning unknown). The fruits are eaten by macaws.
corezeiro, visgueiro (local vernacular). An extract of the bark is used to tan leather in Iquitos, Peru (204). In the Guianas, the barks of various *Parkia* species are attributed medicinal properties (97).

—*Peltogyne* sp. *1
[383]

Tree in *terra firme* forest.
no Waimiri Atroari name given. The wood is used for arrow-heads, and the trunk for house-posts.
pau roxo (local vernacular). The wood of various species of *Peltogyne* is employed for furniture, boat construction, dug-out canoes, tool handles, hunting bows etc. (96).

Sclerolobium eriopetalum Ducke *2
[127]
Tree in *terra firme* forest.
taxi (Waimiri Atroari; *taxi* = a type of ant — a word apparently borrowed from Tupi). The wood is used in house construction, and for fuel. The fruits are eaten by parrots and macaws.

—*Sclerolobium* (?) sp. *3
[228]
Tree in *terra firme* forest.
pykwyky (Waimiri Atroari). The wood is used for fuel.
taxi (Waimiri Atroari). The wood is used for house construction and for fuel. The fruits are eaten by parrots and macaws.

—*Stryphnodendron* sp. *1
[318]
Tree in *terra firme* forest.
no Waimiri Atroari name given. The wood may be used for axe handles and house construction.

Swartzia ingifolia Ducke *1
[431]
Tree in *terra firme* forest.
The middle Amazon valley, mainly around Manaus (76).
kymara (Waimiri Atroari). No uses given.
carrapatinho (local vernacular; = little tick).

Swartzia panacoco (Aublet) Cowan *3
[196]
Tree in *terra firme* forest.
South-eastern Venezuela, the Guianas, and central and eastern Amazonian Brazil (76).
kymara bahnja (Waimiri Atroari; *bahnja* = small). The trunk, which is said to be extremely hard, is used for house-posts.
coração de negro (local vernacular; = black heart). In French Guiana a decoction of the bark is taken to promote sweating, and the wood is said to last at least 60 years buried underground (208). Wayãpi Indian children use the pods as toys (120).

Swartzia ulei Harms *13
[565]
Tree in *terra firme* forest.
Central Amazonia, in the vicinity of Manaus (76).
kyma, kymara (Waimiri Atroari). The trunks are used for house-posts.
jerimum (local vernacular).
Note: It is extremely difficult to distinguish infertile specimens of this species satisfactorily from *S. tomentosa* (Ducke) Ducke.

Tachigali myrmecophila (Ducke) Ducke *5
[607]
Tree in *terra firme* forest.
kanawyba (Waimiri Atroari). The wood was said to be good for fuel, though this is somewhat surprising given that the wood contains skatole, a foul-smelling chemical found in human faeces (7 *fide* 24).
taxi (Waimiri Atroari; *taxi* = a type of ant (*Pseudomyrmex* sp.) which inhabits the leaf rachis). The wood is used for house construction. The fruits are eaten by macaws and parrots.
tachi pitomba (local vernacular). The Kuripako Indians of Colombia use the powdered bark as an antiseptic for ulcerated sores, and rub an infusion of the leaves on aching limbs (252). This species is used as a fever remedy by the Ka'apor of Brazil (23), is attributed medicinal properties by the Tembé (24), and contains tannins (186). The Ka'apor feed the bark to their hunting dogs, apparently to fatten them (23), and in Colombia its ashes are mixed with clay to strengthen pottery (255). This is one of the few species of trees which the Tembé Indians deliberately avoid using as fuel (24), presumably because of its odour.

LINACEAE

Roucheria punctata (Ducke) Ducke
[no voucher retained]
Woody herb (seedling?) in *terra firme* forest.
no Waimiri Atroari name given. This species was said to be used as a cure for snakebite.
azeitona da mata (local vernacular; = forest olive).

LOGANIACEAE

—*Strychnos jobertiana* Baill. vel aff. *1
[559]
Woody liana in *terra firme* forest.
Widespread throughout Amazonia (154).
arawyba (Waimiri Atroari). The fruits are edible, and are eaten by monkeys.
kiriri (Waimiri Atroari; *kiriri* = vine). A fish poison is prepared from the stem.

The Kuripako Indians of Colombia use this species as the primary ingredient of their curare poison (9). The root bark contains tertiary alkaloids (diaboline and jobertine), and is also used by the Kofán of Ecuador to make an arrow poison (154). In Venezuela a perfume is prepared from the leaves to attract the opposite sex (9).

LORANTHACEAE

Genus undetermined GR
[GR2761]
Parasitic plant in seasonally flooded igapó forest.
arawyta xiky ba (Waimiri Atroari; = 'howler monkey's bladder'). Scrapings of the bark are used as plasters for wounds.

The Palikur Indians of French Guiana use several species of Loranthaceae as plasters for broken bones (122), and the Tikuna of Colombia use the crushed leaves of a *Struthanthus* species to stop haemorrhages from small cuts (257). Several species of *Psittacanthus* and *Phoradendron* contain tyramine, which may be responsible for hypertensive responses observed in pharmacological studies (117).

MALPIGHIACEAE

Malpighia punicifolia L.

[no voucher taken]

Shrub cultivated around villages (a very recent introduction). Central America (originally), now cultivated in several neotropical localities.

Waimiri Atroari name not recorded. The fruits are edible.

acerola (local vernacular).

Barbados cherry (English). According to the Creoles of the Antilles, a decoction of the root acts as a diuretic and an infusion of the leaves is an effective gargle for sore throats (191). The wood may be used to manufacture small utensils, the bark for tanning, and the fruits (eaten) are said to be good for liver problems, diarrhoea, dysentery and coughs (280).

MARANTACEAE

Ischnosiphon arouma (Aublet) Koern.

[800] [GR2775]

Herb in damp 'baixada' depressions in the forest.

Panama, Pacific Colombia, northern Venezuela, the Lesser Antilles, Amazonia and the Guianas (14).

baky (Waimiri Atroari). The stems are split vertically, and each strip then split again into an outer and an inner layer. The outer layer is used for fine basketry (when dry), and the inner layer for rough basketry. The leaves may be used for wrapping small packages of food etc. collected in the forest.

arumã (local vernacular). A decoction of the leaves is drunk by the Chácobo Indians of Bolivia to cure diarrhoea, and as an emetic (48). A decoction of the pith of the stem is drunk as an anti-diuretic by the Tiriyó of northern Brazil (63). In Guyana the leaves may be used as an underthatch (97).

Ischnosiphon polyphyllus (Poeppig & Endl.) Koern.

[795]

Tall herb (to 4m) in seasonally flooded depressions in forest and along the river.

The upper Orinoco, the basin of the Rio Negro, and along the Amazon from Tefé to its mouth (14).

warma, warmaxi (Waimiri Atroari). The long flexible stems are scraped, split, and woven into 'bracelets' (*warma*) or may be used for basketry.

MARCGRAVIACEAE

Marcgravia parviflora Rich.

[203]

Woody vine in *terra firme* forest.

French Guiana and Surinam (160), and northern Brazilian Amazonia.

iby (Waimiri Atroari; = silky anteater). Pieces of the bark are chewed to relieve toothache.

Norantea guianensis Aublet

[784]

Woody liana in *terra firme* forest.

Trinidad, Guyana, Surinam and tropical Brazil (160), and Colombia.

kynynyry, saryma nyry (Waimiri Atroari; *saryma nyry* = anteater's tongue). An infusion of the bark is used to bathe babies, or drunk as a cure for diarrhoea.

rabo de arara (local vernacular; = macaw's tail). The Indians of the Apaporis River (Colombia) apply the crushed flowers of this species (which contain flavonoid glycosides (242)) to infected wounds and other abnormal skin conditions (257). Likewise in the Guianas a decoction of the inner bark is used to clean ulcerous sores and cuts, and the pliable rhizomes are used to fix sails to the masts of rowing boats (97).

MELASTOMATACEAE

Maieta guianensis Aublet GR
[GR2711]
Shrub in *terra firme* forest.
Venezuela and Colombia to Amazonian Brazil, Bolivia and Ecuador (293).
iatwa ia (Waimiri Atroari; *iatwa* = *Azteca* ant, *ia* = leaf). The plant is myrmecophilous.

The chewed leaves, applied to wounds, have a haemostatic effect (293).

Miconia navioensis Wurd.
[774]
Small tree in *terra firme* forest.
dinjeny (Waimiri Atroari). The leaves are added to the *karawyra kany* healing brew used at *marba* ceremonies (see p. 43).

Miconia punctata (Desr.) Don ex DC. *1
[632]
Tree in *terra firme* forest.
Southern Mexico and the Greater Antilles to Amazonian Brazil, Bolivia and Ecuador (293).
dinjeny (Waimiri Atroari). The trunks are used for house construction. The fruits are eaten by tapirs.

The Wayãpi Indians of French Guiana prepare a dye from this species (Grenand, pers. comm.).
Note: Many species of *Miconia* are used as sources of (red) dyes in Amazonia, and it was reported by Barbosa Rodrigues (1885) that the Crichaná used melastomataceous dyes to decorate their arrow-heads (although this was not observed amongst the Waimiri Atroari).

Miconia serialis DC. GR
[GR2718]
Small tree in *terra firme* forest.
susupia (Waimiri Atroari). The fruits are edible when they fall.

MELIACEAE

Note: All of the Meliaceae were attributed the name *swakira* by the Waimiri Atroari. However, in a sterile state some of these trees look very similar, and it is possible that there was a degree of confusion over which species may or may not be used to treat stings or to build houses.

Guarea pubescens (Rich.) Adr. Juss. ssp. ***pubescens*** *1
[664]
Tree in *terra firme* forest.

Western and central Amazonia through Amapá State and the Guianas to western Venezuela (196).
swakira (Waimiri Atroari; *swa* = spider) The fruits are eaten by curassows, guans, macaws and parrots.

The Wayãpi Indians of French Guiana regard this species as poisonous, and make arrow-heads from its fresh-cut wood to kill deer (122).

Guarea scabra Adr. Juss. *8
[472]
Tree in *terra firme* forest.
Central and western Amazonia, Pará, Amapá and Cayenne (196).
swakira (Waimiri Atroari). The inner bark is rubbed onto the stings of ponerine ants and scorpions to ease the pain. The wood is used for house construction. The fruits are eaten by curassows, guans, macaws and parrots.

Guarea silvatica C. DC *1
[328]
Tree in *terra firme* forest.
Western and central Amazonia to Guyana, Pará and Maranhão (196).
swakira (Waimiri Atroari). The inner bark is rubbed onto the stings of ponerine ants to ease the pain. The fruits are eaten by curassows, guans, macaws and parrots.

The fruits are eaten by saki monkeys (18).

Guarea trunciflora C. DC. *4
[631]
Tree in *terra firme* forest.
Western and central Amazonia and the Pará/Guyana border (196).
swakira (Waimiri Atroari). The fruits are eaten by curassows, guans, macaws and parrots.

Guarea velutina Adr. Juss. *1
[345]
Tree in *terra firme* forest.
Middle and upper Rio Negro, Roraima and Pará, Brazil (196).
swakira (Waimiri Atroari). As *G. trunciflora.*

—*Trichilia* cf. ***lecointei*** Ducke *3
[591]
Tree in *terra firme* forest.
(Central) and eastern Amazonian Brazil to the Guyana frontier (196) (*T. lecointei*).
swakira (Waimiri Atroari). As *G. trunciflora.*
pracaúba de terra firme (local vernacular). The Tembé Indians of Brazil use *T. lecointei* to dye their hammocks (24).

Trichilia micrantha Benth. *5
[388] [GR2721]
Tree in *terra firme* forest.
Widespread throughout Amazonia to northern Venezuela and Trinidad (196).
swakira (Waimiri Atroari). As *G. trunciflora.*
gitó mirim (local vernacular). The Barasana Indians of Colombia inhale the smoke of the burning leaves to treat pulmonary ailments (257).

—*Trichilia septentrionalis* C. DC *4
[682]

Tree in *terra firme* forest.
Costa Rica and Panama to Venezuela, the Guianas and Amazonia (196).
swakira (Waimiri Atroari). The inner bark is rubbed onto the stings of ponerine ants to ease the pain. The trunks are used for house construction. The fruits are eaten by curassows, guans, macaws and parrots.
no Waimiri Atroari name given. An infusion of the (strong-smelling) bark is drunk to relieve stomach ache.
gitó de terra firme (local vernacular). The Ingano Indians of Putamayo (Colombia) use an infusion of the leaves to treat fever (257).

MENISPERMACEAE

Abuta grandifolia (Mart.) Sandw.
[GR2773]
Small tree or shrub common in *terra firme* forest.
sakraryba, ***iakarara*** *fide* Grenand (Waimiri Atroari; *sakra* = white). The fruits are edible.
cipó abuta (local vernacular). An infusion of the bark is taken by the Tiriyó Indians of Brazil to relieve stomach ache (63). The Wayãpi of French Guiana use a decoction of the grated bark as a dental analgesic, and the Palikur Indians drink it as an aphrodisiac (122). The Siona Sequoia of Ecuador use the wood for house rafters (282). In Ecuador an infusion of the leaves is drunk to alleviate diarrhoea and general body pain, and is used by women for pain relief during childbirth (22). The root bark is a component of the curare arrow poison of the Andoque Indians of Colombia (162), and many other Indian groups also use this species in their arrow or dart poisons (including the Taiwanos, Karijonas and Makunas of Colombia (257)). The Quichua of Ecuador boil the leaves and apply them as a compress to treat sties (157), the Ketchwas (sic) of Ecuador use similar compresses for snake bite (257), and the Siona of Ecuador drink an infusion of the leaves for fever (257). This species contains alkaloids (232). The fruits are eaten by spider monkeys (238).

Abuta sandwithiana Krukoff & Barneby
[no voucher retained]
Woody liana in *terra firme* forest.
xuia pyny kiriri (Waimiri Atroari; = bright-wood or beautiful-wood vine). Pieces of the (yellow) stem are boiled and pounded, and the juice squeezed hot over infected wounds.

The Tenharins Indians use this species to treat wounds, and the Jarawara use it as an ingredient of their arrow poison (label data from INPA herbarium, Manaus). The Karitiana use it as a female contraceptive, and the Wayãpi of French Guiana employ a decoction of the grated bark (which is rich in tetrahydro-protoberberine alkaloids with proven analgesic properties) as a dental analgesic (122).

MONIMIACEAE

Siparuna decipiens (Tul.) A. DC. *2
[412]
Tree in *terra firme* forest.
French Guiana and Surinam (199), and northern Brazil.
ikopy (Waimiri Atroari). The powerful-smelling leaves are eaten, or a decoction of the bark drunk, to cure stomach ache or expel parasites. This species was also classified as *are' bri ieptxi* (medicine for the liver). The trunks are used for house-posts. The fruits are eaten by parrots and macaws.

capitiú (local vernacular; *caa* = vegetation, *pitiú* = foetid (in Tupi)). The leaves of *S. guianensis* Aublet, a species which is widely used for its various medicinal properties and contains alkaloids (122), have similarly been reported to cure stomach ache (8), as have the fruits (257).

MORACEAE

Artocarpus altilis (Park.) Fosb.
[no voucher taken — species noted by Azevedo Lima (1990)]
Tree cultivated around villages.
Widely cultivated in Amazonia; native of the South Pacific Islands.
Waimiri Atroari name not recorded. The fruits are cooked and eaten.
fruta-pão (local vernacular).
breadfruit (English). The male flower spikes may be boiled and eaten. A decoction of the leaves is taken for high blood pressure in the Bahamas, the latex is put on 'liver-spot' skin disease in Jamaica, and in Costa Rica the boiled latex is applied to wounds (187). A decoction of the leaves is also used by the Créoles of French Guiana to lower tension (122). The leaves may be used as animal feed and when young may be chewed to counteract food-poisoning, the wood may be used to make canoes, the bark to manufacture bark-cloth, and the latex to caulk canoes or (diluted) to cure diarrhoea (280). In Samoa this plant is attributed numerous medicinal applications, including the use of a decoction of its root-bark as a febrifuge (276).

Brosimum guianense (Aublet) Huber
[775]
Tree in *terra firme* forest.
Widespread throughout northern South America, Trinidad, southern Panama, Honduras and southern Mexico (35).
piria (Waimiri Atroari; similar names are used throughout the Guianas (Grenand, pers. comm.)). The reddish brown heartwood is highly prized for the fabrication of hunting bows, arrow-points, turtle spears etc.
muirapinima (local vernacular). The wood is used to make furniture in the State of Mato Grosso, Brazil (165), and is used for house-posts in rustic Peruvian architecture (192). It may also be used to make violin bows (137). The fruits are considered edible by the Tukuna (Tikuna) Indians of Colombia (111). The fruits are eaten by saki monkeys (18).

Brosimum parinarioides Ducke ssp. ***parinarioides*** *2
[201B]
Tree in *terra firme* forest.
Lower Amazonian basin to Manaus, and the Guianas (35).
yrmy (Waimiri Atroari). The trunks are used for dug-out canoes. The copious sticky latex may be boiled and added to the latex of *Symphonia globulifera* to make the *manji* resin blocks used in arrow manufacture etc. (see p. 29). The fruits are eaten by tapirs, tortoises and macaws.
amapá roxo (local vernacular). The latex is used in the vicinity of Manaus for healing cuts, and as a cure for general debility (261). It is drunk as a fortifier in the vicinity of Alter do Chão, Pará, Brazil (50), as a cure for internal ulcers by the Palikur Indians of French Guiana (122), and is elsewhere used against lung diseases and sometimes added to balata (35). The fruits are eaten by spider monkeys (238).

—*Brosimum rubescens* Taubert *2
[549]
Tree in *terra firme* forest.

The Amazon basin, the Guianas, northern Panama and around Rio de Janeiro, Brazil (35).

piria sakra (Waimiri Atroari; *sakra* = white). The wood is used for arrow-heads (and bows?).

yrmy (Waimiri Atroari). The trunks are used to make dug-out canoes. The fruits are eaten by tapirs, tortoises and macaws. **Note**: It has been suggested (Grenand and Balée, pers. comm.) that the wood of this species is unsuitable for dug-out canoes. Its identification as *yrmy* may have been a mistake.

pau rainha (local vernacular; = queen tree). The fruits are edible and commercially marketable (203). The wood is attractive, with good elastic properties (233), and is used for furniture and construction in Brazil (165). Several tribes of the eastern Amazon use it for their hunting bows (Balée, pers. comm.).

Clarisia racemosa Ruiz & Pavón

[no voucher taken]

Tree in *terra firme* forest.

Brazil and Peru (159), and Ecuador.

bixwa (Waimiri Atroari). The trunks are used to make dug-out canoes.

guariúba (local vernacular). The wood is used in construction throughout Brazilian Amazonia (166), and possesses particularly good elastic properties (233). In Peru the trunks are used as roof beams in rustic housing (192), and the fruits are used to make an alcoholic beverage (204). The Colorado Indians of Ecuador employ this species as a local anaesthetic for toothache (157). The fruits are eaten by spider monkeys (238) and by capuchin and squirrel monkeys (271).

Ficus guianensis Desv. ex Ham. *1

[629]

Strangler in *terra firme* forest.

Northern Brazil (to Goiás), Venezuela, the Guianas, Trinidad and Grenada (37).

kypý (Waimiri Atroari). The latex is used to cure eye infections, infected skin sores and wounds. The fruits are eaten by small birds. Note that the same Waimiri Atroari name was also given to two species of *Clusia* (Guttiferae).

mata-pau (local vernacular; = kill-tree). The latex of other species of *Ficus* have likewise been attributed healing properties in Colombia (257). The Kuripako Indians of Colombia take the latex for stomach pains, and in the environs of Leticia (Colombia) the Indians regard it as calmative for digestion (257). The fruits are eaten by spider monkeys (238).

Helianthostylis sprucei Baill. *6

[627]

Tree in *terra firme* forest.

Amazonian Brazil and Colombia (35).

pirywyxixi (Waimiri Atroari). The latex from the bark, which was said to cause itching, is applied to infected wounds to promote healing. It was also attributed the medicinal property *kory ieptxi* (*kory* = pus). A leafy branch is beaten against the lower legs to ward off snakebite after one has dreamed of snakes. The fruits are eaten by *meiyra* (an unidentified animal).

The bark and fruits contain alkaloids (232).

Helicostylis elegans (Macbr.) C.C. Berg *1

[136]

Tree in *terra firme* forest.

Upper Amazon basin (Peru and Brazil) (35).

axi (Waimiri Atroari). The fruits are eaten by tortoises, and by spider and howler monkeys.
inharé (local vernacular). The wood is used for house construction in Peru (202).

Helicostylis tomentosa (Poeppig & Endl.) Rusby *2
[126]
Tree in *terra firme* forest.
Northern South America east of the Andes, and coastal Brazil north of Rio de Janeiro (35).
axi (Waimiri Atroari). The latex is applied to wounds to assist healing. The fruits are eaten by tortoises, spider and howler monkeys.
inharé (local vernacular). The latex is said to be toxic by the Indians of the Leticia region of Colombia (254). Extracts of the inner bark of this species, whose latex is used as a hallucinogen by the Indians and Bush Negroes of Surinam, produced effects similar to those of *Cannabis* on the physiology of laboratory rats (55). The fruits are eaten by the Chácobo Indians of Bolivia (48), and by the Dení of Brazil (215). The Andoque of Colombia roast the seeds and eat them (162). Latex of *H. scabra* (Macbr.) C.C. Berg is used by the Puinave Indians of Colombia to prevent infection of abrasions (249). The fruits are eaten by spider monkeys (238), brocket deer and tapir (120).

Maquira calophylla (Poeppig & Endl.) C.C. Berg *1
[611]
Tree in *terra firme* forest.
Amazonian Brazil, Peru and Venezuela (35).
xipieni (Waimiri Atroari). The fruits are edible, and are eaten by tortoises and spider monkeys.

The bark contains maquiroside, which is a potentially chemotherapeutic cytotoxic cardiac glycoside (241), and furanocoumarins (240). In the environs of Leticia (Colombia), the Indians regard the latex from the bark of this species as caustic and toxic (257).

Naucleopsis amara Ducke *6
[582]
Tree in *terra firme* forest.
Amazonian Brazil and Colombia (35).
asy (Waimiri Atroari). The copious sap from the bark is applied externally to fungal infections and infected wounds. The fruits are edible, and are eaten by monkeys, and the leaves are eaten by sloths.

Naucleopsis mello-barretoi (Standl.) C.C. Berg *5
[249]
Tree in *terra firme* forest.
Brazilian Amazonia, and the coast from Bahia to Rio de Janeiro (35).
mawakyny (Waimiri Atroari). The latex is applied to infected wounds. The fruits are eaten by monkeys.

The latex contains a complex mixture of cardiotonic glycosides, principally antiarin (46). It is used by the Makú Indians of northern Brazil as an arrow poison (215). The fruits are eaten in the vicinity of Iquitos, Peru (279).

Naucleopsis ternstroemiiflora (Mildbr.) C.C. Berg *5
[153]
Tree in *terra firme* forest.

Amazonian Brazil and Peru (35).
mawakyny (Waimiri Atroari). The fruits are edible, and are eaten by monkeys.

Perebea xanthochyma Karsten GR
[GR2743]
Small tree in *terra firme* forest.
Southern Costa Rica to Brazil (35).
xipieni (Waimiri Atroari). The fruits are edible.

Pourouma bicolor Mart. ssp. ***bicolor*** *5
[366]
Tree in *terra firme* forest.
Throughout the Amazon basin and the Guianas, extending to northern Colombia and Bolivia (38).
tarawy, ikemi iba (Waimiri Atroari; *ikemi iba* = trumpeter's eye). The fruits are edible. The wood is used in house construction. The fruits are eaten by monkeys, trumpeters and curassows.

Pourouma cf. ***bicolor*** ssp. ***digitata*** (Trécul) C.C. Berg & van Heusden *1
[61]
Tree in *terra firme* forest.
Northeast Amazonia and the Guianas (38).
tarawy (Waimiri Atroari). See *P. bicolor* for uses etc.

Pourouma cucura Standl. & Cuatrec. *2
[692]
Tree in *terra firme* forest.
Widespread in northern Amazonia, extending to eastern Venezuela (38).
tarawy, ikemi iba (Waimiri Atroari). See *P. bicolor* for uses etc.
embaúbarana branca (local vernacular). An infusion of the bark is rubbed into the skin by the Taiwano Indians of Colombia to ease rheumatic pains (252).

Pourouma guianensis Aublet ssp. ***guianensis***
[772]
Tree in *terra firme* forest.
Amazonia (throughout), the Guianas and eastern Brazil (38).
kawyxixi (Waimiri Atroari). Pieces of the inner bark (which smell of 'Vick' ointment) are rubbed onto the forehead to relieve cold symptoms, or an infusion of the same is prepared and drunk.
embaúba benguê (local vernacular; *benguê* = embrocation). The fruits are eaten by the Chácobo Indians of Bolivia (48), the Tembé of Brazil (24) and the Wayãpi of French Guiana (Grenand, pers. comm.).

Pourouma minor Benoist *1
[595]
Tree in *terra firme* forest.
Northern South America to Central America (Nicaragua) (38).
kyry' (Waimiri Atroari). The sawdust causes skin irritation, and smoke from the burning wood causes irritation of the lungs. The fruits are eaten by spider monkeys.
embaúbarana (local vernacular). The fruits are apparently eaten in the vicinity of Iquitos, Peru (279), although this is the only species of *Pourouma* **not** eaten by the Wayãpi Indians of French Guiana (120). The fruits are eaten by spider monkeys (238) and by capuchins (271).

Pourouma myrmecophila Ducke
[no voucher taken]
Treelet in *terra firme* forest.
Colombia (Amazónas), Peru (Loreto) and Brazil (Amazonas) (38).
iatwa pyrma (Waimiri Atroari; *iatwa* = a type of ant). The fruits are edible.

Pourouma ovata Trécul *4
[691]
Tree in *terra firme* forest.
Venezuela (Bolívar), Peru (Loreto) and Brazil (Acre, Amazonas and western Pará) (38).
kyry' (Waimiri Atroari). The sawdust causes skin irritation, and smoke from the burning wood causes irritation of the lungs. The fruits are eaten by howler monkeys.
embaúbarana (local vernacular; = false *Cecropia*).

Pourouma tomentosa Miq. ssp. ***tomentosa*** *1
[216]
Tree in *terra firme* forest.
The northwest of the upper Amazon basin (38).
kukwiri (Waimiri Atroari). The fruits are edible, and are eaten by monkeys.
imbaúba do vinho (local vernacular; *vinho* = wine).

***Pseudolmedia laevigata* Trécul** *2
[456] [GR2728]
Tree in *terra firme* forest.
Northern South America (35).
pama (Waimiri Atroari). The fruits are edible, and are eaten by monkeys, toucans, guans and parrots.

The Indians of the Apaporis river, Colombia, drink an infusion of the young twigs to stimulate the appetite (254), and the Andoke Indians believe that the latex helps to rejuvenate the blood (257).

Pseudolmedia laevis (Ruiz & Pavón) Macbr. *1
[763]
Northern South America, east of the Andes (35).
pama (Waimiri Atroari). The fruits are edible, and are eaten by monkeys, toucans, guans and parrots.
pama (local vernacular). The fruits are eaten by howler monkeys, capuchin monkeys and squirrel monkeys (271).

Sorocea guilleminiana Gaudich. *3
[474]
Tree in *terra firme* forest.
Widespread throughout the Amazon basin, and eastern Brazil to Minas Gerais and Rio de Janeiro (36).
mare iepry (Waimiri Atroari; = sieve tree). The flexible twigs are used to make the frames for sieves.

This species is said to possess medicinal properties by the Chácobo Indians of Bolivia (48). The Kayapó of Brazil plant it as a source of edible fruits and medicine (13).

MUSACEAE

Musa cultivars
[no voucher taken]
Giant herb cultivated in swidden sites and around villages.
Widely cultivated throughout the tropics; native of Asia.
wyky (Waimiri Atroari). The fruits are eaten, or made into a beverage and drunk. At least eight varieties are recognized: *keryry* ('inaja'); *kybyma* ('maça'); *waiwai wykry* ('branca', exchanged with Waiwai Indians); *kaiy* ('bai'); *matyra* ('prata'); *kyrakyny* ('pacovã'); *kwarakyny* and *xyiwa wykyry*.
banana, pacovã (local vernacular).
banana, plantain (English). According to the Wayãpi Indians of French Guiana, the sap bled from the trunk possesses antiseptic properties (122). It has been used in Jamaica to treat diarrhoea (292). The Island Caribs of Dominica apply the juice squeezed from the inflorescence buds to ulcers (136). It is said in Brazil that the eating of unripe bananas cures stomach ulcers (WM pers. obs.). Other uses of bananas, which include extraction of starch from waste green fruit, use of the foliage as cattle fodder, production of weak rope from the leaf sheath fibres and the use of corms, shoots, male buds and pseudostem hearts as vegetables, are summarized by Stover & Simmonds (1987).

MYRISTICACEAE

Note: The majority of the Myristicaceae were attributed the same Waimiri Atroari name: *warxi*. Although *dia* (*Osteophloem platyspermum*), *kapkwany* (*Iryanthera tricornis*) and *tapany iepy* (*Virola michelii*) were readily identified as separate entities, the distinction between those named as *waryry*, *w'axi* and *warxi* appeared to be less straightforward — or more confused.

—*Iryanthera juruensis* Warb. *4
[165]
Tree in *terra firme* forest.
Amazonian Brazil and Peru (265), Bolivia, Ecuador and Colombia.
warxi (Waimiri Atroari). The trunks are used in house construction. The fruits are eaten by toucans, parrots, howler monkeys, guans and tinamous.
wytxiry basekry (Waimiri Atroari; = sloth's claw). Used as a vermifuge for children.
ucuúba punã (local vernacular). The Chácobo Indians of Bolivia drink a decoction of the leaves as an emetic (48), and the Waorani of Ecuador use the resin from the bark to cure fungal infections of the skin (81). The Tembé of Brazil attribute this species medicinal properties (24). The fruits are used as fish bait in Colombia (9), and are eaten by the Bora of Peru (102). Oil is extracted from the seeds (8). The fruits are eaten by woolly monkeys (144).

Iryanthera laevis Markgraf *1
[720]
Tree in *terra firme* forest.
Amazonian Brazil and adjacent Peru (265), Bolivia, Colombia and Guyana.
wa'xi (Waimiri Atroari). Resin from the bark is applied to wounds to prevent infection and accelerate healing. Designated as *kirita ieptxi* (medicine for unknown complaint). The wood is used in house construction. The fruits are eaten by toucans and toucanets.
ucuúba punã (local vernacular). In Guyana an infusion of the bark of certain *Iryanthera* species is employed as an antidote for poison (97). A decoction of the leaves of this species is taken as an emetic by the Chácobo Indians of Bolivia (48). The Andoque Indians of Colombia eat the aril of the fruits (162), which contain flavolignoids (108).

Iryanthera paraensis Huber *9

[195]

Tree in *terra firme* forest.

Amazonian Brazil and adjacent Peru and Guyana (265), and Ecuador.

warxi (Waimiri Atroari). The wood is used for house construction. The fruits are eaten by toucans, parrots, tinamous and guans.

ucuúba (local vernacular). The Waorani Indians of Ecuador apply the bark resin to fungal infections of the skin and mite infestations (81), as do the Indians of the upper Apaporis River in Colombia (257). The Dení of Brazil boil the fruits and eat the aril (215), as do the inhabitants of the Iquitos region, Peru (279).

Iryanthera tricornis Ducke *2

[264]

Tree in *terra firme* forest.

Amazonian Brazil (265), and Colombia.

kapkwany, kapkweni (Waimiri Atroari; *kapy* = sky).

The wood is used to make paddles and stools, and is employed in house construction. The fruits are eaten by toucans, spider monkeys and toucanets.

ucuúba punã (local vernacular). The Indians of the Soratama region of Colombia use a decoction of this species to treat infected cuts and burns (256). The Indians of the Rio Purús, Brazil, use its wood to make blowpipes (257).

Osteophloem platyspermum (A. DC.) Warb. *4

[449]

Tree in *terra firme* forest.

Amazonian Brazil and adjacent Peru and Colombia (265).

dia (Waimiri Atroari). The seed is roasted and eaten. When the trees are fruiting, it is said that the time has come to begin hunting spider monkeys. According to one informant the sap may be taken as a febrifuge. The fruits are eaten by spider and saki monkeys.

ucuúba branca, ucuúba chorona (local vernacular). The Makú Indians of Brazil drink the sap to cure coughs and colds, and around Manaus the leaves are burned and the smoke inhaled to relieve asthma (255). The Kuripako Indians of Colombia burn the dried leaves in their homes after a fever or cold (257). The latex is used to heal wounds in Colombia (1), and the Ingano Indians of Mocoa use scrapings of the soft inner bark for the same purpose (257). The Makunas of the Río Piraparaná apply it externally to relieve the pain of the stings of ponerine ants (257). The wood of this species contains isoflavonoids (51).

Virola calophylla Warb. *1

[588]

Tree in *terra firme* forest.

Widespread throughout the Amazon basin (234).

warxi (Waimiri Atroari). The wood is used in house construction. The fruits are eaten by toucans, parrots, guans and tinamous.

ucuúba vermelha (local vernacular). The Makú Indians of Brazil drink an infusion of the bark to cure malaria (215), whereas in Peru this infusion is taken to treat belly and bladder disorders (257). The Waorani of Ecuador apply the bark resin to fungal and scabies infections of the skin (81). The bark resin is used to prepare a hallucinogenic snuff by the Indians of Colombia (245). The alkaloids (tryptamine derivatives) present in this species are at their most concentrated in the leaves (5). In Loreto, Peru, this species is attributed minor commercial value (204).

Virola elongata (Benth.) Warb. *1

[687]

Tree in *terra firme* forest.
Amazonia, the Guianas and Venezuela. Also found in Panama (234).
warxi (Waimiri Atroari). The wood is used in house construction. The fruits are eaten by toucans, parrots, guans and tinamous.
ucuúba punã (local vernacular). The resin from the bark is used by the Makú Indians of Brazil to prepare a hallucinogenic snuff (215), and likewise by the Yanomamo (52). There are also reports of its use as an arrow poison, but analysis of the resin has revealed no obviously toxic compounds (173). The Witoto of Colombia use this species to treat fungal infections of the skin (256) and analyses of the bark have revealed the presence of eusiderin, a neolignan thought to possess antifungal properties (172). The Barasana Indians of Colombia apply a decoction of the leaves and twigs to swollen joints (257). In Loreto, Peru, this species is attributed minor commercial value (204).

Virola michelii Heckel *3
[265]
Tree in *terra firme* forest.
Central and northern Amazonian Brazil and the Guianas (234).
tapany iepy (Waimiri Atroari; = hard tree). The resin from the bark is applied to aching teeth to alleviate pain.
ucuúba preta (local vernacular). This species is used by the Ka'apor Indians of Brazil to cure sore throats (23). Resin from the bark of *V. peruviana* (A. DC.) A.C. Smith is used to relieve toothache in the Leticia region of Colombia (256), that of *V. sebifera* Aublet is used in French Guiana (208), and the use of a myristicaceous bark resin to ease teething pains has also been reported from the Paumarí Indians of Brazil (223). The aril of this species contains 47% fat (201). The fruits are eaten by toucans and cacique birds (120).

—*Virola pavonis* (A. DC.) A.C. Smith *1
[116]
Tree in *terra firme* forest.
Peru, Colombia, Venezuela and the Brazilian states of Amazonas, Pará and Roraima (234).
waryry (Waimiri Atroari). Resin from the bark is applied to the bites of ponerine ants to deaden the pain (see *Osteophloem*). The fruits are eaten by spider monkeys, toucans and guans.
ucuúba branco do baixo (local vernacular). The Bora Indians of Peru use the resin from the bark of this species as a hallucinogen (258). The bark contains eusiderin, a neolignan thought to possess antifungal properties (114). The wood is used in construction (234).

—*Virola theiodora* (Benth.) Warb. *3
[94]
Tree in *terra firme* forest.
Note: Some claim this species to be synonymous with *V. elongata*.
warxi, waryry (Waimiri Atroari). Resin from the bark is applied to wounds to assist healing. The fruits are eaten by spider monkeys, toucans and guans.
ucuúba (local vermacular). The leaves are aromatic and serve as a substitute for tea (256). Resin from the bark is used as an arrow poison and as a hallucinogenic snuff by the Waiká (Yanomama) Indians of Brazil (215). Resin is mixed with the ashes of *Gustavia poeppigiana* Berg ex Mart. (Lecythidaceae) and ingested orally as a hallucinogen by the Witoto of Colombia (246).

MYRTACEAE

Note: There was considerable discrepancy in the naming of the Myrtaceae collected in the hectare study plot.

—*Eugenia* sp. *3
[243]
Tree in *terra firme* forest.
kymaky sehery (Waimiri Atroari; *kymaky* = leafcutter ant). The fruits are edible, and are eaten by capuchin monkeys.
kabkaby (Waimiri Atroari). The fruits are edible.
kwapitxeni (Waimiri Atroari). The trunk may be used for house-posts. The fruits are eaten by tortoises.

—*Eugenia/Myrciaria* sp. *1
[544]
Tree in *terra firme* forest.
tikxiryra (Waimiri Atroari). The wood is used to make arrow-heads. According to one informant the fruits are edible. The fruits are eaten by tortoises.

Marliera spruceana Berg GR
[GR1781]
Small tree in damp 'baixada' depression in forest.
Amazonian Peru and Brazil (174), and Colombia.
imieni (Waimiri Atroari). The fruits are edible.

The Taiwano Indians of Colombia drink a hot decoction of the leaves of this species to clear the throat of congestion (257).

Myrcia guianensis (Aublet) DC. GR
[GR2714]
Tree in seasonally flooded igapó forest.
Widespread in South America east of the Andes (174).
warda (Waimiri Atroari). The fruits are edible.
Note: According to Baines (pers. comm.), a plant called *warda* grows amongst the rocks in river beds and is used to prepare a fragrant wash for new-born babies. This may be *Myrciaria dubia* (Kunth) McVaugh.

—cf. ***Myrcia minutiflora*** Sagot *2
[171]
Tree in *terra firme* forest.
Northern Brazilian Amazonia and the Guianas (174).
kyrywaky ie (Waimiri Atroari; = parrot's tooth). The wood is used to make arrow-heads, and the trunk may be used for house-posts.

The fruits of *M. minutiflora* are eaten by the Wayãpi Indians of French Guiana (120).

Psidium guajava L.
[no voucher taken]
Tree cultivated around villages.
Cultivated throughout the tropics; native of the New World.
Waimiri Atroari name not recorded. The fruits are eaten.
goiaba (local vernacular).
guava (English). An infusion of the bark, roots and leaves is taken by the Island Caribs of Dominica to cure diarrhoea (136). Similar uses are reported from the Tikuna Indians of Colombia (257), the Wayãpi of French Guiana (122) and the people of

Morobe Province, Papua New Guinea (288). The Creoles of the Antilles use a decoction of the leaves as a gargle for sore throats (191). The people of Papua New Guinea chew them with salt to relieve colds or flu (288). The leaves may also be chewed to relieve toothache, and the roots may be eaten in a soup (87). This species is rich in tannins (62), and possesses antibacterial properties (281). The fruits are particularly rich in vitamin C (53). A summary of the multifarious uses to which this plant is put in Africa is given by Watt & Breyer-Brandwijk (1962).

Syzygium malaccense (L.) Merr. & Perry
[no voucher taken]
Tree cultivated around villages (a very recent introduction).
Cultivated throughout the tropics; native of the Malay Peninsula (170).
Waimiri Atroari name not recorded. The fruits are edible.
jambo, jambeiro (local vernacular).
Malay apple (English). In Papua New Guinea the sap from the leaves is used to treat severe coughs, and the scraped bark is applied to sores. An infusion of the bark is also taken for diarrhoea and stomach ache (288). In Samoa it is likewise used to treat stomach upsets, as well as for coughs, sore throat, and stomach bloat after childbirth (276). The wood may be used for construction (280).

Genus undetermined *4
[157]
Tree in *terra firme* forest.
tapyrma pyny (Waimiri Atroari; = black wood). The wood is used to make arrow-heads. The trunks are used for house-posts. The fruits are eaten by parrots.

The use of myrtaceous woods for arrow-heads is widespread in Amazonia and the Guianas (Grenand, pers. comm.).

NYCTAGINACEAE

Neea cf. ***altissima*** Poeppig & Endl. *3
[760]
Tree in *terra firme* forest.
sasyma iepy (Waimiri Atroari). The fruits are eaten by capuchin and saki monkeys, and by parrots.
João mole (local vernacular; = soft John). The leaves of several species of *Neea* are chewed in Colombia to blacken the teeth (257).

Neea sp. *1
[194]
Tree in *terra firme* forest.
sasyma iepy (Waimiri Atroari). The wood is used for smoking meat and fish. The fruits are eaten by capuchin and saki monkeys, and by parrots.
João mole (local vernacular). The Carib Indians of Guyana prepare a body paint from ripe *Neea* fruits (97).

Neea sp. *1
[652]
Tree in *terra firme* forest.
As species above.

OLACACEAE

Heisteria barbata Cuatrec. *3
[642]
Tree in *terra firme* forest.
Amazonian Colombia, Venezuela, Peru and Brazil (263).
tyny' (Waimiri Atroari). The trunks are used for house-posts, and make good firewood. The fruits are eaten by howler and capuchin monkeys.
Note: the same name was given to *Bocoa viridiflora* (Leguminosae).
castanha de cutia (local vernacular; = agouti's chestnut).

—*Heisteria laxiflora* Engl. *1
[267]
Tree in *terra firme* forest.
Amazonian Brazil (263).
pakia ihy (?) (Waimiri Atroari; *pakia* = collared peccary). No uses given.

Heisteria spruceana Engl. GR
[GR2757]
Tree or shrub in igapó forest.
Amazonian Colombia, Venezuela, Peru and Brazil (263).
paxiy'ra (Waimiri Atroari). No uses given.

The Indians of the Apaporis and Piraparaná rivers of Colombia use an infusion of the leaves and branches to relieve swelling and pain in sprained joints (254).

Minquartia guianensis Aublet *2
[563]
Tree in *terra firme* forest.
Central America, the Guianas and Amazonia (263).
tedewydewyra (Waimiri Atroari). The trunks are used for house-posts.
acariquara (local vernacular; = acari-fish burrow in Tupi). According to the Wayãpi Indians of French Guiana, the fruits are edible (120). The wood of this species is used widely in Amazonia for telephone and electricity posts etc. It also produces a reasonable substitute for sandalwood oil (263). The Desana Indians of Brazil eat the caterpillars which feed on the leaves (228). In Ecuador the inner bark is applied to skin abrasions and infections, to lessen pain and promote healing (22). A polyacetylene with cytotoxic properties, named minquartynoic acid, has been isolated from this species (177). The Quijos Quichua of Ecuador drink an infusion of the inner bark as an anthelminthic, and the Canelos Quichua take it for stomach ache (177). The Waoranis and Ketchwas of Ecuador use the bark as a fish poison (257). In French Guiana the wood shavings are boiled in water to produce a black dye for cotton (208). The fruits are eaten by spider monkeys (238). The seeds are eaten by saki monkeys (18), and are dipersed by pacas and agoutis (141).

—*Ptychopetalum olacoides* Benth. *3
[612]
Tree in *terra firme* forest.
The Guianas and Amazonia (263).
tybyxiryra (Waimiri Atroari). The trunks are used for house-posts.
tybyiryra (Waimiri Atroari). An infusion of the bark may be drunk to cure stomach ache.
muirapuama (local vernacular). A reputedly aphrodisiac drug, prepared chiefly from the roots of the young plants, is exported from Manaus and Belém (263). It constitutes one of the ingredients of a medicine used by the *caboclos* of the lower Amazon to treat

rheumatism (10). A decoction of the roots is also said to prevent baldness and paralysis (163). The bark, of which a decoction is used by the Wayãpi Indians of French Guiana as a fortifying bath for children (120), contains lipophilic esters, lupeol and behenic acid (17 *fide* 263).

ORCHIDACEAE

Brassia bidens Lindley
[488]
Epiphyte in seasonally flooded igapó forest.
Venezuela, Guyana, Brazil and Peru (103).
Waimiri Atroari name not recorded. The pseudobulb is heated in water, and its sticky mucilage applied to fungal infections of the skin (which appear as white blotches on the shoulders).

PALMAE

Astrocaryum aculeatum Meyer vel aff. *1
[no voucher taken]
Tree in *terra firme* forest.
Brazilian Amazonia and the Guianas (110).
sykma', sykuma (Waimiri Atroari). The nutritious flesh around the seeds is eaten raw. It contains 47.2% fat, 5.5% protein, 6.8% carbohydrate and is rich in Vitamin A (4). The seeds are broken open and the endosperm is also eaten. The (yellow) pinnae of unopened fronds are used to weave mats and fans. The fruits are eaten by peccaries, pacas and agoutis.
tucumã (local vernacular).
black palm, pimpler palm (English). The 'wood' is used for bows and for arrow-heads by the Chácobo Indians of Bolivia, and the leaves are used in basketry (48). The Xiriana (Yanomama) Indians of northern Brazil regard the 'palm-heart' as edible (12), and the Kayapó extract salt from this species (211). The epicarp may be used to smoke-cure rubber (74). The Puinave Indians of Colombia extract the oil from the seeds and rub it warm onto the chest to relieve pulmonary problems (257).

Astrocaryum jauari Mart. GR
[no voucher taken]
Tree in igapó forest.
Brazil, Peru, Colombia and the Guianas (110).
iawara (Waimiri Atroari). These trees are important for locating fish, which feed from their fallen fruits and can be shot with arrows as they do so.
jauarí (local vernacular). The pinnae may be woven into mats, and provide a source of strong fibre. The epidermis of the petiole may be woven into mats and manioc-squeezers, and the mesocarp of the fruit contains an edible oil (205). Brazilian ladies of the upper Amazon use the seeds to make heads for their lace-making bobbins (284).

Astrocaryum sp.
[no voucher taken]
Apparently acaulescent spiny palm in *terra firme* forest.
kwirba (Waimiri Atroari; == 'our fan' — possibly not the actual name of the plant). The pinnae of unopened fronds are used to weave mats and fans.
Note: This may be *Astrocaryum acaule* Mart., which was found by Grenand (pers. comm.) in damp depressions in the forest under the name *xibididi*, and whose fruits were said

to be eaten by cutias and pacas. According to Wallace (1853), the pulp of these fruits is edible but not particularly pleasant.

Bactris gasipaes Kunth
[no voucher taken]
Tree cultivated around villages.
Cultivated throughout northern South America and Central America; native of the Amazon.
merepy (Waimiri Atroari). The fruits are boiled and eaten.
pupunha (local vernacular).
peach palm (English). The wood is sometimes used by the Chácobo Indians of Bolivia to make hunting bows and arrow-heads (48). The Tikuna Indians from the Leticia region (Colombia) extract a green dye from the young leaves for decorating bark-cloth (257). The Tucano Indians of Brazil prepare an alcoholic beverage from the fruits (113). The Yurumanqui Indians extract salt from the spadix, and the Borucas use this species to make their fire-sticks (87). In pre-Colombian times the roots were used as a vermicide, the male flowers were used for flavouring food, and the seeds were eaten (68). The 'palm-heart' is edible (87). Some Brazilian tribes use the spines from the stem as needles for tattooing the skin with soot (284). The fruits are eaten by parrots and macaws.

Bactris humilis (Wallace) Burret SR
Small palm in *terra firme* forest.
Widespread in tropical South America (48).
xitomy (Waimiri Atroari).The fruits are edible.
tucum marajá (local vernacular). The Chácobo Indians of Bolivia drink a decoction of the fruits to treat stomach ache (48).

Bactris sp. GR
[no voucher taken]
Small understorey palm in *terra firme* forest.
piriti (Waimiri Atroari). The hard gracile stems are sometimes used to make arrow shafts.

Cocos nucifera L.
[no voucher taken]
Tree cultivated around villages.
Cultivated throughout the tropics; probably native of Malesia.
kyky (Waimiri Atroari; corruption of local vernacular). The endosperm of the seed is edible. This species is a recent introduction.
coco da bahia (local vernacular).
coconut (English). This species is used by the Créoles of French Guiana to treat boils, and contains phenolic compounds with antifungal properties (122). The Tikuna Indians of Colombia drink an infusion of the scorched leaf bases to prevent impending miscarriages (111). In Papua New Guinea the young fronds are heated over the fire, and the extract is applied to fresh knife or axe wounds (288). The other multifarious applications of this species are summarized elsewhere (e.g. 612, 987).

Euterpe controversa Barb. Rodr. SR
Palm in *terra firme* forest.
Central Brazilian Amazonia (110).
wesi mepry (Waimiri Atroari). The palm-heart is edible.
açai chumbinho (local vernacular). An astringent beverage may be made from the fruits, and the stem may be used for house construction (74).

Euterpe precatoria Mart. *1
[802]
Tree occasionally found in *terra firme* forest, but more often in damp 'baixada' depressions.
Peru, Colombia, Venezuela, Brazil, Trinidad and the Guianas (110).
wesi, manaka (Waimiri Atroari; the first name is possibly borrowed from Tupi). The fruits are soaked in warm water and eaten, or made into a nutritious beverage. The fruits are eaten by toucans and other birds.
açai (local vernacular). A decoction of the leaves is taken by the Chácobo Indians of Bolivia to cure chest pains, and the leaves are used for thatch (48). The 'palm-heart' is edible. The trunks are used in house construction by the Andoque Indians of Colombia (162) and by the Xiriana (Yanomama) of Brazil, who also prepare a black body paint from its fruits and use the inflorescences as brooms (12). The dry fruits contain 4% oil (201), and are eaten by spider monkeys (238) and by capuchin monkeys (145).

Geonoma deversa (Poit.) Kunth SR
Small palm in *terra firme* forest.
Widespread throughout Amazonia, the Guianas and Central America (110).
kaneia (Waimiri Atroari). The leaves are sometimes used for thatching.
ubim bravo (local vernacular). The Xiriana Indians of Brazil use the stem of this species to make arrow-points and the leaves to make temporary baskets. Although they occasionally use its leaves for thatching, they are not the first choice for this application and are not very long-lasting (12).

Iriartella setigera (Mart.) H.A. Wendl.
[no voucher taken]
Treelet in *terra firme* forest.
Southeast Colombia, southern Venezuela, Guyana & western Brazil (132).
irwa (Waimiri Atroari). The trunks are split into thin flexible strips, which are used to make fish-traps (*maxa*, see Fig. 6). The split trunks may also be woven together with *ahnjaty* (*Heteropsis*) roots to make the doors of the houses.
paxiubinha (local vernacular). The Kuripako Indians of Colombia make blow-guns from the stem (31). This species yields a valuable fibre, and the stem may be used to make a traditional Indian musical instrument known as the 'chicuta' (74). In the State of Amazonas, Brazil, an infusion of the roots is used as a stimulating bath (40). The palm-heart is said to be edible (54).

Jessenia bataua (Mart.) Burret
[no voucher taken]
Tree in damp 'baixada' depressions in the forest.
Brazil, Peru and Colombia (110).
patwa (Waimiri Atroari; possibly borrowed from Tupi). The fruits are soaked in warm water and eaten, or made into a nutritious beverage. The palm-heart is edible. The seeds may be dried and burned, and salt extracted from the ashes. Disposable carrying baskets for game or vegetable products may rapidly be woven from the leaves, which are also used to make temporary shelters. The stiff fibres (*maxekyba*) from the bases of old leaves are used as ceremonial arrows at the *marba* festivities.
patauá (local vernacular). The fruits contain approximately 7.4% oil, whose flavour is similar to that of olive oil (201). It is employed for its anti-tubercular properties by the Witoto Indians of Colombia, and is mixed with almond oil and used as a purgative and as a cure for chronic cough in the region of Tingo Maria, Peru. It is said to promote hair growth, and is used by the Indians of Peru to affix *Genipa* dye to the skin. An infusion of the seedlings is used as a remedy for snakebite by the Bora Indians of

Peru, who also eat the grubs (*Rhynchophorus palmarum*) which can be found in the felled trunks (31). The Waorani Indians of Ecuador prepare a medicine for headaches, diarrhoea, worms and other stomach complaints from the adventitious roots (81), and use the leaves for thatch. The immature inflorescence is edible and nutritious. The Cubeo Indians of Colombia use the spines of the leaf sheath as blow-gun darts (they are also good for lighting fires), and the Guahibo use its wood to make their hunting bows and arrow-points (31). It may also be used for making walking sticks (74). The Xiriana (Yanomama) of northern Brazil make heavy war-clubs from the wood of this palm, and prepare a blue body dye from its fruits (12). In Guiana the kernels are sometimes eaten by Indians in times of hunger; the Waiwai grate them and mix them with manioc flour (97).

Manicaria sp. (probably *M. martiana* Burret)
[no voucher taken]
Short-trunked tree not found in *terra firme* forest.
wy' (Waimiri Atroari). The leaves are used for thatching.
bussu (local vernacular). There is considerable confusion over the taxonomy of the genus *Manicaria*. *M. saccifera* Gaertn., which is considered by some to be its only member (286), has been attributed a variety of useful properties (other than thatching) (e.g. 153).

Mauritia aculeata Kunth GR
[no voucher taken]
Tree in areas of poor drainage on sandy soils.
Venezuela, Brazil and Colombia (110).
karbana (Waimiri Atroari). The leaves are used for thatching. The petiole is sometimes used to make arrow shafts.
caranaí (local vernacular). A pleasant beverage may be prepared from the fruits (284).

Mauritia carana Wallace
[no voucher taken]
Tree in white-sand 'caatinga' forest.
Venezuela, Brazil and Colombia (110).
karbana (Waimiri Atroari). The leaves are used for thatching.
caranã (local vernacular). A thatch made from the leaves of this plant is said to last from eight to 10 years (284).

Mauritia flexuosa L.
[no voucher taken]
Tree common in damp 'baixada' depressions in the forest.
Widespread throughout Amazonia and the Guianas, and Trinidad (110).
mixi (Waimiri Atroari). A delicious beverage is made from the fruits (which are soaked first to soften them). The fibrous epithelia are stripped from the pinnae of unopened fronds, and twisted into a string from which hammocks are made. Hunting hides may be set up amongst the trees of this species, since the fallen fruits attract many game animals. In the past the seeds were dried thoroughly over a fire, burned, and salt was extracted from the ashes. The fruits are eaten by tapirs, agoutis and other animals.
buriti (local vernacular). The leaves are used as thatch by many Amazon peoples. A sago-like starch is prepared from the trunk by the Warao Indians of the Orinoco (248), and a wine may be prepared from the sap. This sap is very rich in sugar, and in Pará (Brazil) is used as a source of the same. The outer shell and mesocarp contain 12% oil, and the seed and kernel 5% (201). The Xiriana (Yanomama) of northern Brazil extract this oil by boiling the fruits, and spread it on manioc bread to eat (12). The 'palm-heart' is said to be sweet and delicious (193), and the fruits contain

extremely high levels of Vitamin A (4). In Colombia the Guahibo Indians eat the grubs which can be found in the fallen trunks of this species (30), and in the State of Amazonas (Brazil) an infusion of the seeds is taken for intestinal ills (40). The leaves contain saponins and glycosides, and are used by the Warao Indians to treat coughs, pertussis and fever (287). Canoes may be made from the hollowed trunk (74). The Guajajara Indians of northeast Brazil weave their manioc-squeezers from the fibres of the outer portion of the leaf petioles (32), and the Makuxi of Guyana (and many other tribes) make sandals from the leaf bases (239). In the Guianas, starch from the trunk is said to be good for dysentery and diarrhoea (97). The fruits and flowers are eaten by tortoises (188). The fruits are eaten by white-lipped peccaries, collared peccaries and tapirs (47).

Maximiliana maripa (Corr. Serr.) Drude
Tree in *terra firme* forest. This species may also be managed (possibly planted) in swidden sites (Grenand, pers. comm.).
The Guianas and northern Brazilian Amazonia (110).
mapia (Waimiri Atroari). The fruits (and seeds) are edible, and the leaves are sometimes used for thatching (though not for the round-houses).
inajá (local vernacular). The palm-heart is edible and delicious, and the leaves are used for thatching by a number of Indian groups. The Xiriana of Brazil use them to weave walls for their houses (12), and in the Guianas they are used to make fans and baskets (239). The fruits may be burned to smoke-cure rubber, and the stem may be used in house construction (73). The Arawak Indians of the Pomeroon River (Guyana) used the fruit pedicels for making fire. The Waorani Indians of Ecuador make their blow-gun darts from the split petioles, and drink an infusion of the fruits to treat colds (81). The ashes of the leaves are regarded by the Kubeo Indians of Kuduyarí as an antidote to poisoning by *Mayna toxica* R.E. Schultes (Flacourtiaceae) (247). The woody spathes may be used as receptacles for cooking food or boiling water, or as cradles (284). Chácobo children use them as toys (48). The Makuxi Indians of northern Brazil use the oil from the seeds as ear drops (239). In the State of Roraima the beetle grubs ('gonga'), which often inhabit fallen fruits, are either eaten or used as bait for pacu fish (WM pers. obs.). The fruits are eaten by monkeys and some birds (284).

Oenocarpus bacaba Mart. *24
[803]
Tree common in *terra firme* forest.
Brazil, Peru, Colombia and the Guianas (110).
kymy (Waimiri Atroari; = hand). The fruits are soaked in warm water and eaten, or made into a nutritious beverage. The palm-heart is edible. The stems of young individuals are boiled, and the hot juice squeezed over infected wounds to heal them. The fruits are eaten by guans, tinamous, toucans, trumpeters and curassows.
bacaba (local vernacular; = oily fruit in Tupi). The Créoles of French Guiana use the 'palm-heart', prepared in a number of ways, to treat open wounds (122). The immature inflorescence is edible and nutritious (31). A light yellow oil, similar to that of the olive, may be extracted from the fruits (74). The Xiriana (Yanomama) of northern Brazil use the wood to make their hunting bows and arrow-points (12), and it may also be used for walking sticks, umbrellas, spears and a sacred Indian musical instrument known as the 'pirey' (74). The leaves are used for basketry by the Brazilian Pareci and Nhambiquara Indians (74), and they also serve for thatching. The fruits are eaten by spider monkeys (238).

Socratea exorrhiza (Mart.) H.A. Wendl.
[no voucher taken]
Tree generally found in damp 'baixada' depressions.
Nicaragua south to Bolivia, and east through Venezuela, the Guianas and Brazil (132).

paxiy' (Waimiri Atroari; probably borrowed from Tupi). The leaves of immature individuals, when crushed in water, produce a gelatinous substance which serves as soap. The seeds were split and woven into the *kwyie* aprons which were originally worn by Waimiri Atroari women. The trunk is occasionally used for construction, although this practice may have been adopted from outsiders (see p. 40).

paxiúba (local vernacular). Since the wood can easily be split into planks with an axe, it is much used for construction in Amazonia by Indians and settlers alike. A decoction of the bark is drunk by the Chácobo Indians of Bolivia as a remedy for fever, and in the past the spiny stilt-roots were used as manioc-graters (48). The Waorani Indians of Ecuador make peccary spears from the wood (81), and the Xiriana (Yanomama) of Brazil occasionally use it to make their hunting bows (12). The Indians of the Vaupés river (Colombia) use the trunk to make ceremonial trumpets for their initiation rites (247), and make drills for piercing ornamental quartz stones from the wood (239). The Palikur of French Guiana regard the fruits as poisonous, and use plasters of the grated stem to heal the umbilicus of the new-born (122). In the State of Amazonas, Brazil, an infusion of the roots is used as a restorative bath (40). It is said that eating the 'palm-heart' of this species can result in sexual impotence (L. Coelho, pers. comm.). This species is sometimes used to kill cockroaches (132). The Maiongon (Yekuana) Indians of Brazil use the exudate from the prop roots as a constituent of their curare arrow poison (207), and the Andoque Indians of Colombia (162) and the Waorani of Ecuador (81) eat the (hard and bitter) fruits. Salt may be extracted from this species (74). The leaves may be used for thatching, and there are records of the use of an unspecified part of the plant as an abortifacient (132). The fruits are eaten by spider monkeys (238), and by birds (284).

PASSIFLORACEAE

Passiflora nitida Kunth GR

[GR2766]

Vine in secondary vegetation.

West coast of Colombia, through the Guianas to Peru and Goyaz, Brazil. Also in Panama (cultivated?) (150).

meretxa, myretxi (Waimiri Atroari). The fruits are edible.

maracujá do mato (local vernacular).

wild passion fruit, bell apple (English). In Peru an infusion of the leaves is drunk as a general tonic (202). The Makuna Indians of Colombia consider the eating of the fruits to be beneficial to elderly people with failing eyesight (257). They are sold in the market in Iquitos, Peru (279).

PIPERACEAE

Piper consanguineum Kunth

[776]

Herb in *terra firme* forest.

Surinam, French Guiana and northern Brazil (295).

tykymawariri (Waimiri Atroari). The leaves are macerated with a little water or saliva, and the juice applied to infected cuts or snakebite. The leaves are a constituent of the *karawyri kany* healing brew used at *marba* ceremonies (see p. 43). If a dog is made to sniff the crushed roots, his hunting proficiency will be improved.

Note: The Waimiri Atroari claimed (in Portuguese) that this species is used to cure 'ferida brava' (literally 'fierce wound'), which could be interpreted either as infected wounds or as leishmaniasis.

The Chácobo Indians of Bolivia use the stem exudate to heal wounds (48). Several species of *Piper* are applied to snakebite by the Cuna Indians of the western Amazon (104).

Piper sp.
[777]
Herb in *terra firme* forest.
kiamky se (Waimiri Atroari; *kiamky* = toucan). This name was applied to several species of *Piper*. The leaves are combined with a little water or saliva, and the juice squeezed onto arrow wounds to heal them.

Several species of *Piper* are used by the Chácobo Indians of Bolivia to heal cuts and sores (48). The Wayãpi of French Guiana use *P. submelanostictum* DC. for the same purpose (122).

POLYGALACEAE

Moutabea sp. *1
[193]
Woody liana in *terra firme* forest.
mapyskana (Waimiri Atroari). The fruits are edible, and are eaten by monkeys.

POLYGONACEAE

—***Coccoloba*** sp. *1
[519]
Woody liana in *terra firme* forest.
kiriri (Waimiri Atroari; = vine). No uses given.

QUIINACEAE

Lacunaria jenmanii (Oliver) Ducke *1
[743]
Tree in *terra firme* forest.
Guyana, Amazonian Brazil and Peru (161).
iamy tyrwe (Waimiri Atroari; *iamy* = tinamou). The fruits are edible, and are eaten by small monkeys (*kyxiri*) and curassows.
moela de mutum (local vernacular; = curassow's crop).

Quiina negrensis A.C. Smith *1
[141]
Tree in *terra firme* forest.
mare'sowpry (Waimiri Atroari; *mare'* = guan). The trunks are used for house-posts. The fruits are eaten by guans.

RUBIACEAE

Duroia saccifera Benth. ex Hook. *1
[181]
Tree in *terra firme* forest.
Brazilian Amazonia and adjacent Venezuela (268).

wiriapry (Waimiri Atroari; *wiria* = tayra). The fruits are edible, and are eaten by tayras.
cabeça de urubu (local vernacular; = vulture's head). In Colombia the seeds are said to be toxic if ingested (254).

Ferdinandusa sp. *2
[633]
Tree in *terra firme* forest.
wyrawyra, waruwara (Waimiri Atroari). The wood burns extremely fiercely, even when fresh-cut, and is well suited for camping in the forest and for cremations. The trunks are used for house-posts.

Genipa sp.
[no voucher taken]
Tree cultivated around villages.
Waimiri Atroari name not recorded. A black body-paint is prepared from the fruits.
jenipapo (local vernacular).
genip (English). Although this plant was not seen, it is likely to be *G. americana* L., whose use as a body paint is widespread and for which many other uses (food, medicine etc.) have been recorded.

Palicourea quadrifolia (Rudge) Steyerm.
[294]
Shrub or treelet in *terra firme* forest.
Guyana, French Guiana and Amazonian Brazil (267).
paxyhra (Waimiri Atroari). An infusion of the bitter-tasting bark is drunk to cure diarrhoea or cough. A little of the bark is chewed into a paste and applied to foot wounds to prevent infection. The fruits are eaten by tapirs.

This species was attributed medicinal properties by the Chácobo Indians of Bolivia (48). Another species of *Palicourea, P. corymbifera* (M. Arg.) Standl., is used to treat coughs by the Desano Indians of the lower Vaupés, Colombia (257).

—***Psychotria*** (?) sp. *1
[710]
Tree in *terra firme* forest.
no Waimiri Atroari name given. The fruits are eaten by tapirs.

RUTACEAE

Citrus aurantifolia (Christm.) Swingle
[no voucher taken]
Tree cultivated around villages.
Cultivated throughout the tropics; native of Asia.
arexa (Waimiri Atroari). A drink made from the fruits is recognized as a cold cure.
limão (local vernacular).
lime (English). The Wayãpi Indians of French Guiana rub the mashed leaves on the forehead for headaches. The juice is used as an antiseptic by the Créoles of the Antilles (191). The rind of immature fruit, beaten into a pulp, may be applied to the eyelids to cure sore eyes, and the fruit may also be used to cleanse laundry (124). In Colombia the Tikuna Indians take a mixture of the juice with sugar and aspirin as a febrifuge and abortifacient, and claim that a decoction of the roots taken during menstruation acts as a contraceptive (111). In Samoa the juice is gargled for sore throats, and the leaves are used to treat filariasis (276).

Citrus sinensis (L.) Osbeck
[no voucher taken]
Tree cultivated around villages.
Cultivated throughout the tropics; native of Asia.
arexa (Waimiri Atroari). The fruits are edible.
laranja (local vernacular).
orange (English). The Créoles and Palikur Indians of French Guiana drink an infusion of the leaves as a cold cure (122). Oil from the seeds and rind is used in soap manufacture etc. (170).

SAPINDACEAE

Cardiospermum halicacabum L. var. ***microcarpum*** Kunth
[no voucher taken]
Herbaceous climber/scrambler in secondary vegetation.
Pantropical.
Waimiri Atroari name not recorded. The seeds are occasionally used as beads.

Cardiospermum halicacabum is used as a fish poison in the Guianas (278 *fide* 3). Arm-bands of these seeds are thought by the Indians of the Apaporis basin (Colombia) to ward off snakebite (257). The oil from the seeds is reputed to possess insect repellent and antifeedant properties (257).

Matayba inelegans Radlk. *2
[125]
Tree in *terra firme* forest.
Northern Amazonia and British Guiana (92).
kysiwyra (Waimiri Atroari). If a man steps on the fruits he will not find tortoises in the forest. The fruits are eaten by parrots and macaws. Note that this species was referred to by the same Waimiri Atroari name as were several *Sloanea* species (Elaeocarpaceae).
breu de tucano (local vernacular; = toucan's resin)

SAPOTACEAE

Chrysophyllum prieurii A. DC. *2
[230]
Tree in *terra firme* forest.
Central Brazilian Amazon to the Guianas, Venezuela, Colombia, Panama and Peru (195).
mehe (Waimiri Atroari). The fruits are edible and highly sought after. The fruits are eaten by monkeys.
abiurana vermelha (local vernacular; = red false abiu). The wood is used for construction around Manaus (195). The fruits are eaten by peccaries (195), and by howler monkeys (202).

Ecclinusa ramiflora Mart. *1
[532
Tree in *terra firme* forest.
Amazonia, Pacific Colombia and Brazilian Atlantic forest (195).
asawara (Waimiri Atroari). The fruits are edible, and are eaten by capuchin and saki monkeys.
coquirana brava (local vernacular).

Manilkara bidentata (A. DC.) Chev. ssp. ***surinamensis*** (Miq.) Penn. *1
[389]
Tree in *terra firme* forest.
Caribbean islands, Guianas and northern Amazonia (195).
parteki (Waimiri Atroari). The fruits are edible. A dab of the sticky latex from the bark, capped with a piece of leaf, is used to draw out parasitic botfly larvae from the skin. The wood is used for house construction. The fruits are eaten by spider and capuchin monkeys.
maparajuba*, *maçaranduba vermelha (local vernacular).
balata (English). The latex was formerly exported as chicle (195). In Peru the wood is used for parquet flooring (202). In the Guianas the bark is used as an emetic, and the latex and a decoction of the bark are ingredients of a dysentery cure (97). The Wayãpi Indians of French Guiana prepare a glue for arrow-making from the latex (120).

Manilkara sp.
[no voucher taken]
Tree common in white-sand 'caatinga' forest.
parta (Waimiri Atroari). The trunks are favoured for use as house-posts.

In the Guianas, the Warrau Indians use a species of *Manilkara* to make bark canoes (239).

Micropholis guyanensis (A. DC.) Pierre ssp. ***guyanensis***
[788]
Tree in *terra firme* forest.
Widely distributed throughout northern South America, southern Central America and the Caribbean (195).
ararba (Waimiri Atroari). The fruits are edible, and are eaten by monkeys.
chicle bravo (local vernacular; = wild chicle). The Chácobo Indians of Bolivia drink a decoction of the leaves for rheumatism (48). The fruits are eaten by spider monkeys (238).

Pouteria anomala (Pires) Penn. *1
[75]
Tree in *terra firme* forest.
Southern Venezuela and Amazonian Brazil (195).
karamxi (Waimiri Atroari). The fruits are edible, and are eaten by monkeys.
abiurana roxa (local vernacular; = purple false abiu).

—***Pouteria caimito*** (Ruiz & Pavón) Radlk. *1
[630]
Tree in *terra firme* forest, and cultivated around the village. Costa Rica to central Amazonia, and coastal Brazilian forest to Rio de Janeiro (195).
tysyhyna ipe (Waimiri Atroari; wild). ***wairaky*** (cultivated). The fruits are edible, and are eaten by monkeys.
abiu*, *abiurana ferro (local vernacular). This species is widely cultivated as a fruit tree in northern South America. The latex is used to cure eye sties by the inhabitants of Alter do Chão, Pará, Brazil (50), and the fruits have been found to contain a broad spectrum of triterpenes (194). The Witoto Indians of the Río Caquetá (Colombia) macerate and toast the young leaves, and apply them to wounds as a disinfectant (257). An extract of the seed or of the bark can be used to treat diarrhoea and intermittent fevers, and the freshly squeezed juice of the fruits may be taken for pulmonary or bronchial disorders (206). The wood is durable and resistant (206).

Pouteria cuspidata(?) (A. DC.) Baehni *1
[683]
Tree in *terra firme* forest.
Pacific Colombia, southern Venezuela and Amazonia (195).
sakiripia, iekyne'byri ipe (Waimiri Atroari). The fruits are edible (highly regarded). The trunks are used for house-posts. The fruits are eaten by macaws, parrots and capuchin monkeys.
abiurana (local vernacular).

Pouteria engleri Eyma *2
[280]
Large tree in *terra firme* forest.
Guianas to central Amazonian Brazil (195).
kiry (Waimiri Atroari). The fruits are edible, and are eaten by white-lipped peccaries and monkeys.

The wood is suitable for general construction purposes (96), and in Loreto (Peru) this species is attributed major commercial value (204).

—**Pouteria glomerata** (Miq.) Radlk. ssp. ***stylosa*** (Pierre) Penn. *5
[766]
Tree in *terra firme* forest.
Costa Rica, Panama, Venezuela and Amazonian Brazil to Maranhão State (195).
karamija, kaieki iemry ba (Waimiri Atroari; = deer's testicle). The fruits are edible. The trunks may be used for house-posts. The fruits are eaten by monkeys.

Pouteria guianensis Aublet *3
[744]
Tree in *terra firme* forest.
Trinidad, western Colombia to the Guianas, Amazonian Brazil and Peru and the Brazilian coastal forests (195).
karba (Waimiri Atroari). The fruits are edible (highly regarded). The fruits are eaten by monkeys.
abiurana abiu (local vernacular). The latex was exploited as balata in the past (195). The wood is suitable for heavy construction (96). The fruits are eaten by spider monkeys (238).

—**Pouteria reticulata** (Engl.) Eyma ssp. ***reticulata*** *1
[187]
Tree in *terra firme* forest.
Mexico through Central America to the Guianas, Amazonia, and the Atlantic Brazilian forests to Rio de Janeiro (195).
kawyxi (Waimiri Atroari). The fruits are edible, and are eaten by howler monkeys.
twadka, tuwadka (Waimiri Atroari).
abiurana cascuda (local vernacular). The wood is suitable for general construction purposes (96).

Pouteria retinervis Penn. *1
[543]
Tree in *terra firme* forest.
French Guiana and northern Amazonian Brazil (195).
twadka, tuwadka (Waimiri Atroari)
The fruits are eaten by monkeys and peccaries.
abiurana grande (local vernacular).

Pouteria vernicosa Penn. *1
[172]
Tree in *terra firme* forest.
Peruvian and western Brazilian Amazon (195).
tapyrma ipe (Waimiri Atroari; *tapyrma* = black, *ipe* = fruit). The fruits are edible, and are eaten by monkeys.
abiurana floribunda (local vernacular).

Pouteria sp. *1
[655]
Tree in *terra firme* forest.
wa'dakyny (Waimiri Atroari). The fruits are said to cause madness if eaten. The fruits are eaten by spider monkeys.

—*Pouteria* sp. *1
[657]
Tree in *terra firme* forest.
twadka, tuwadka (Waimiri Atroari).
mapyskyny (Waimiri Atroari). The fruits are eaten by monkeys.

Sarcaulus brasiliensis (A. DC.) Eyma *2
[503]
Tree in *terra firme* forest.
Costa Rica and Panama to Amazonia and coastal Brazil (195).
kyxiyrakry (Waimiri Atroari; *kyxiy* = saki). The fruits are eaten by capuchin, saki and squirrel monkeys.
abiurana batinga (local vernacular). The fruits are edible (195).

Genus undetermined *7
[241]
Tree in *terra firme* forest.
twadka, tuwadka (Waimiri Atroari). The trunks of small individuals were used for house-posts in the past. The fruits are said to have a foetid odour. The fruits are eaten by monkeys and peccaries.

—Genus undetermined *1
[351]
Tree in *terra firme* forest.
no Waimiri Atroari name given.

Genus undetermined *7
[399]
Tree in *terra firme* forest.
twadka, tuwadka (Waimiri Atroari). See [241] (above) for uses etc.

Genus undetermined *1
[441]
Tree in *terra firme* forest.
karba sakra (Waimiri Atroari; *sakra* = white). The trunks are used for house-posts. The fruits are eaten by monkeys.

—Genus undetermined *1
[477]
Tree in *terra firme* forest.
twadka, tuwadka (Waimiri Atroari). See [241] (above) for uses etc.

Genus undetermined *1
[585]
Tree in *terra firme* forest.
ararba (Waimiri Atroari). The fruits are edible, and are eaten by monkeys.

Genus undetermined *1
[681]
Tree in *terra firme* forest.
batwa (Waimiri Atroari). The fruits are edible, and are eaten by monkeys.

—Genus undetermined *1
[no voucher (lost)]
Tree in *terra firme* forest.
ararba bahnja (Waimiri Atroari; *bahnja* = small). The fruits are edible, and are eaten by monkeys.
pirikeny (Waimiri Atroari). No uses given.
Note: Since there is no surviving voucher specimen of this tree, the possibility that it belongs to one of the species previously listed cannot be dismissed.

SIMAROUBACEAE

Picrolemma sprucei Hook. f.
[no voucher retained]
Woody herb (seedling?) in *terra firme* forest.
prepy (Waimiri Atroari). The stem is chewed to relieve toothache.

This species is used by the Jamamadí Indians of Brazil as a constituent of their curare arrow poison, and an infusion of the bark is drunk by the Paumarí Indians as a cure for malaria (218). The Bora Indians of Peru use it to relieve skin irritations (83).

SOLANACEAE

Capsicum spp.
[no voucher taken]
Shrub cultivated around villages.
warasy (Waimiri Atroari). The fruits are occasionally eaten. However, the Waimiri Atroari do not share the fondness for pimenta which many other Indian peoples possess.
pimenta (local vernacular & English).

STERCULIACEAE

Sterculia pruriens (Aublet) K. Schum. *1
[668]
Tree in *terra firme* forest.
tykexy sowypry (Waimiri Atroari). Strips of the fibrous inner bark are used as carrying straps for babies and panniers. The fruits are eaten by guans.
tacacazeiro, xixá (local vernacular). The Takanos of Colombia use the oil from the seeds to cure 'mange' (255), and the Taiwanos apply the ashes of the bark (mixed with a fat or an oil) to pimples and sores of the scalp (257). The Créoles of French Guiana add wood-ash

of this species to tobacco snuff to increase its narcotic effect (122), and The Ka'apor Indians of Brazil use the wood to build their looms (23). The fruits are eaten by toucans and monkeys (26).

Theobroma grandiflorum K. Schum.
[no voucher taken]
Tree cultivated around villages; a recent introduction.
Widely cultivated throughout Amazonia; native of Pará State, Brazil (78).
aka' (Waimiri Atroari). The aromatic pulp around the seeds may be eaten, or used to prepare a refreshing drink.
cupuaçu (local vernacular). Ice cream made from these fruits is now sold in the USA.

Theobroma subincanum Mart. *2
[307]
Tree in *terra firme* forest.
Widespread through Amazonia (from Pará westwards), Venezuela (upper Orinoco) and French Guiana (78).
aka' (Waimiri Atroari). The fruits are edible. The trunks are used for house-posts, and are good for fuel. The fruits are eaten by capuchin monkeys.
cupuí (local vernacular). Ashes of the bark are added to tobacco snuff by the Dení and Jamamadí Indians of Brazil (215), and to *Virola* snuff by Indians in Colombia (245).

Theobroma sylvestre Mart. *5
[235] [GR2714]
Tree in *terra firme* forest.
Along the Amazon and the lower reaches of its tributaries (78).
srykwaba (Waimiri Atroari). The cauliflorous fruits are edible, and are eaten by howler and capuchin monkeys, and by *iarky* (unidentified monkey).
cacauí (local vernacular). The wood is used in construction in the vicinity of Manaus (261).

TILIACEAE

Apeiba echinata Gaertner *1
[316]
Tree in *terra firme* forest.
Amazonia and the Guianas (48).
kyme' (Waimiri Atroari; = comb). The spiny fruits were used in the past as combs, and the wood is good for fuel. The fruits are eaten by parrots and macaws.
pente de macaco (local vernacular; = monkey's comb). The Ka'apor Indians of Brazil regard this tree as an ornamental (23). The Chácobo of Bolivia use strips of the inner bark for basketry when other materials are unavailable, and originally used the wood for stools (48). The flowers and fruits are eaten by spider monkeys (238).

Lueheopsis rosea (Ducke) Burret *2
[288]
Tree in *terra firme* forest.
aka'ra (Waimiri Atroari). The wood is used to make paddles, and is good for fuel. The fruits are eaten by parrots.
açoita cavalo (local vernacular; = horse-whip).

VIOLACEAE

Rinorea racemosa (Mart.) O. Kuntze *1
[753]
Tree in *terra firme* forest.
Widespread over South America, but unknown from the Guianas (130).
kybahy ihy*, *kybaxy tynry (Waimiri Atroari). The wood may be used for making arrowheads, and is good for fuel. The fruits are eaten by tapirs and tortoises.
branquinha (local vernacular; = small white one). The Bora Indians of Peru use the trunk for house construction, make charcoal from the wood, and attribute (unspecified) medicinal properties to this plant (102).

VOCHYSIACEAE

—*Erisma fuscum* Ducke *1
[569]
Tree in *terra firme* forest.
no Waimiri Atroari name given.
quarubarana (local vernacular).

Qualea sp. *2
[65]
Tree in *terra firme* forest.
maty (Waimiri Atroari). Slabs of the bark are added to the *karawyri kany* healing brew used at *marba* ceremonies (see p. 43). Scrapings of the cambium may be applied externally for relief of localized aches and pains. The fruits are eaten by parrots, parakeets and *kyxiri* monkeys.
mandioqueira (local vernacular; = manioc tree). In Colombia the bark of *Qualea homosepala* Ducke is considered a potent styptic, and used to treat machete wounds. It possesses haemostatic effects and assists healing (257).

ZINGIBERACEAE

Renealmia floribunda K. Schum.
[796]
Herb in *terra firme* forest.
Northern South America, including Trinidad (169).
ne'neia (Waimiri Atroari). An infusion of the (aromatic) leaves is used for bathing babies (*pitxeni kipytxiwy*, medicinal value uncertain).

The components of the essential oils of this species have been reviewed (168). A decoction of the rhizome of a *Renealmia* species is used to induce sterility by the Colorado Indians of Ecuador (157).

FAMILY UNDETERMINED

—*Genus undetermined* *1
[468]
Tree in *terra firme* forest.
no Waimiri Atroari name given. The fruits are edible.

USEFUL PLANTS NEITHER COLLECTED NOR IDENTIFIED TO FAMILY LEVEL

asadan iepy (Tree, *terra firme* forest).
Juice from the inner bark is squeezed onto wounds.

bawyka kiriri (Liana, *terra firme* forest).
The red sap is applied to snakebite.

kaxiri (Tree, *terra firme* forest).
A fish poison is prepared from the bark.

kymyba (Tree, *terra firme* forest).
Scrapings of the inner bark are macerated with a little water or saliva, and applied externally for localised pain (e.g. to the forehead for headaches).

kyrykwy (Tree, *terra firme* forest).
The bark is applied externally as a treatment for pain.

kwaxie (Tree, *terra firme* forest).
The inner bark is rubbed on the skin (medical application uncertain).

ma'xi (Tree, *terra firme* forest).
The bark provides a powerful fish poison.

tamxa iepy (Vine, *terra firme* forest).
A medicine for menstrual disorders is prepared from the stem.

tynakyrykwa, anakyrykwa, wanyny (Tree, *terra firme* forest).
The fruits are edible. This is probably *Macoubea* sp. (Apocynaceae).

wateki (Tree, *terra firme* forest).
An infusion of the bark possesses medicinal properties. This species serves as an insect repellent.

Note: The Wayana and Tiriyó Indians use the name 'wataki' for a species of *Geissospermum* (Apocynaceae), a genus which is of considerable medicinal importance in French Guiana and Surinam (Grenand, pers. comm).

wudyi
This species may be used to treat the sting of the sting-ray.

IMPORTANCE OF THE FOREST TO THE WAIMIRI ATROARI

In one hectare of *terra firme* forest, 200 species of tree (dbh greater than or equal to 10cm) and 14 species of vine/strangler were recorded. Of these, 175 (81%) were attributed useful properties by the Waimiri Atroari informants. As a result of occasional differences of opinion between informants, five of those species were (on different occasions) given more than one Waimiri Atroari name, of which at least one was associated with no useful property. If these five are discounted as useful species, one arrives at an overall species utilization figure of 79%. In addition, 141 of those species (66%) were reported to provide indirect benefit to the Indians by furnishing food sources for their game animals. If one were to include those species (and also those which may be used as firewood) as 'useful', as they have been in similar studies with other Indian groups (see p. 1), then the overall percentage of useful species in the plot would be 92%.

The number of species allocated to each of the principal use categories discussed previously is shown in Table 4. In addition, as a demonstration of the broader potential value of this hectare of forest, utilization percentages which include the *additional* applications gleaned from the literature (see Table 3) are also presented. Considering such uses in conjunction with those assigned to the plants by the Waimiri Atroari, the overall percentage of species in the hectare with useful properties assigned to them would be 86%.

Table 4
Utilization of one hectare of forest by the Waimiri Atroari

	Species (W-A only)	% spp. (W-A only)	% spp. (including literature search)
Trees: (200 spp. total).			
Food	54	27	38
Medicine	30	15	31
Construction	65	32	45
Technology/Other	63	31	42
Commercial	0	0	
Vines/Strangler: (14 spp. total).			
Food	7	50	57
Medicine	3	21	29
Construction	0	0	0
Technology/Other	3	21	35
Commercial	0	0	

It is evident from these data that the Waimiri Atroari inhabit a forest of extraordinary floristic diversity; the number of tree species recorded in one hectare of forest is one of the highest so far encountered in Brazilian Amazonia (see Campbell *et al.*, 1986). It is also evident that their highly developed knowledge of the trees' properties and of inter-specific relationships of flora and fauna enables them to exploit those resources, renewably, to a remarkable extent. This has resulted in the utilization of an extremely wide variety of plant and animal species.

In fact the trees studied in this survey constitute but a small proportion of the plant resources available to the Waimiri Atroari. Likewise the botanical information recorded here represents but a fraction of the knowledge which they possess. The available time, and a desire to make this study comparable with previous ones, limited the minimum diameter of the trees recorded in the survey to 10cm, but had this been lower many more species would have been recorded. For instance, a reduction of the minimum survey diameter from 10cm to 5cm in 80m^2 of a similar forest studied by Prance *et al.* (1976), raised the species count from 41 to 72. Similarly, had it been possible to survey more than one hectare of forest (see the species/area curve in Fig. 8), the species total would have been higher still.

Many tree species of importance to the Waimiri Atroari were inevitably missed by this survey. *Caryocar villosum* (Caryocaraceae) and *Bertholletia excelsa* (Lecythidaceae), for example, were common in the area and are both important food species for the Indians, but they were not present within the plot. Furthermore, only a brief study was made of the uses of herbs, shrubs, epiphytes

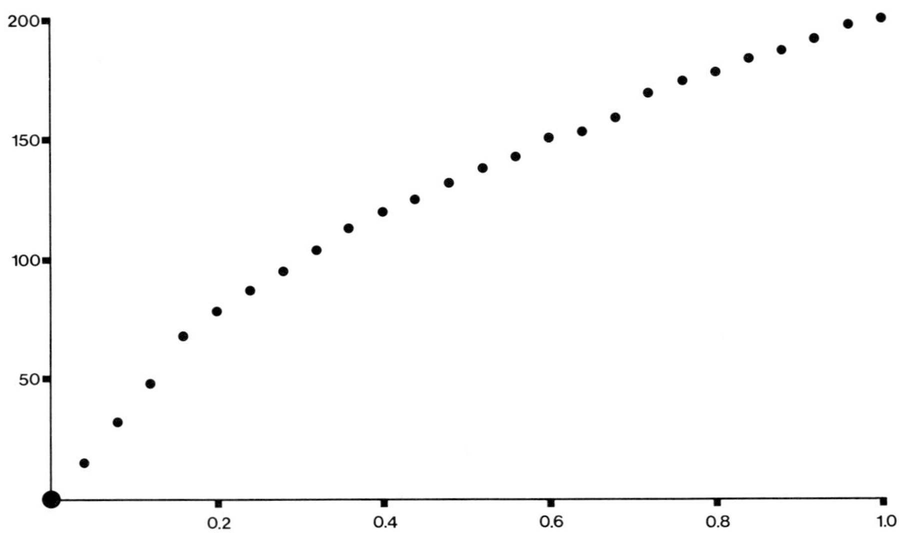

Figure 8. Species/area curve for one hectare of *terra firme* forest close to the village of Maré.

and climbers, and yet it is these plants (and not the trees) which appear to furnish the Waimiri Atroari with their main source of traditional medicines. In addition, the degree to which the other vegetation types are utilized: the igapó, the caatinga forests, the secondary vegetation, etc., remains unknown.

Table 5
Comparison of the use of the trees of one hectare of forest by five tribes of Amazonian Indians*

Tribe	WA	KA	TE	CH	PA
Number of species	200	99	119	94	70
Utilization (%)	79	77	61	79	49
Food (%)	27	34	22	40	34
Medicine (%)	15	21	11	35	11
Construction (%)	32	20	30	17	3
Technology/Other (%)	31	27	25	19	4
Commerce (%)	0	2	5	1	4

WA:- Waimiri Atroari; KA:- Ka'apor; TE:- Tembé; CH:- Chácobo; PA:- Panare

* incorporating the data of Balée & Boom, as presented by Prance *et al.* (1987)

A great deal still remains to be discovered about Waimiri Atroari use of plants, but the data presented in this paper are sufficient to demonstrate unequivocally that the Waimiri Atroari are heavily dependent upon the *terra firme* forest for every aspect of their physical, cultural and spiritual lives. In order to retain an adequate supply of the plants which they need (many of which are present in very low densities on account of the diversity of the flora), and of the animals which they hunt, extensive tracts of forest are clearly required. Considering the results of similar surveys with other Indian groups (see Table 5), which demonstrated comparable levels of dependence upon and utilization of the forest, it could reasonably be supposed that this is the case for all forest Indians.

It is interesting to compare the traditional use and management of forest resources by the Waimiri Atroari with agricultural projects recently installed within their reserve. In 1983 it was decided that cattle pasture should be created on their behalf, ostensibly as a supplementary source of protein. The funding for this project was provided by the Paranapanema mining company, whose other food supplementation projects included the provision of frozen chickens and beef. The Waimiri Atroari, traditionally a hunting tribe, not only have no experience or tradition of animal husbandry, but also appear to have very little interest in it. To maintain the project since its inception, professional 'gaucho' ranchers have had to be brought in by Programa Waimiri Atroari to fell and burn the pastures, and to look after the animals (Plate 6f).

The incompatibility of Amazonian *terra firme* forest with the ranching of cattle has been learnt by long and bitter experience in Brazil. It has finally been accepted that without the unrealistic support of tax incentives, and without the annual destruction of vast tracts of forest, this practice is not a viable one (Buschbacher, 1986). The implementation of such a project thus threatens not only to lower the traditional esteem of the Indians for their primary resource, but to undermine the arguments for their need for its support without providing any workable alternative or even any significant benefit.

The Waimiri Atroari, unlike many other Amazonian peoples, have been allocated a large reserve which was officially recognized at the highest level possible under Brazilian law at the time of this study. Their population is now increasing, and with it their demands upon the forest's resources and the importance of maintaining the integrity of its boundaries. In the recent past, however, vast tracts of land have been wrested from the Waimiri Atroari for mineral extraction and hydroelectric power generation. And such is the commercial interest in the valuable minerals which underlie their forests* that one cannot dismiss the possibility of further losses occurring. Given the degree to which the Waimiri Atroari depend upon their natural resources, any such losses would be disastrous for their future.

ACKNOWLEDGEMENTS

This research would not have been possible without the help and support of a great many people, principally the Waimiri Atroari Indians without whose knowledge, generosity, patience, friendship and eagerness to teach, nothing could have been achieved. The project was initiated by Rodrigo de Azevedo Lima, coordinator of Programa Waimiri Atroari's ecological scheme, with the aim of providing a base of information by which the external forces acting upon the Indians' environment, and the consequential effects upon the Indians themselves, could be assessed. He arranged the necessary permits and provided much valuable advice and support.

The fieldwork was largely funded and backed by the Programa Waimiri Atroari, which provided transport, food, accommodation and equipment. Some of its employees offered invaluable assistance and much appreciated kindness. Botanical collecting equipment was lent by Bruce Nelson (Projeto Flora Amazonica), the Royal Botanic Garden, Edinburgh, and Jill Thompson (University of Stirling). W.M.'s travel and living expenses during 1989 were largely met by the Royal Geographical Society and the Peter Nathan Cultural Trust, respectively. Matt Murphy (Sherkin Island Marine Station) provided invaluable advice in the initial phases of preparation. The World Wide Fund for

* On 24 June 1989 an agreement (Termo de Compromisso 001/89 — since repudiated) was signed which allocated the mining rights for the entire Waimiri Atroari reserve to the Paranapanema mining group in return for monthly royalty payments to the Indians.

Nature's 'Biological Dynamics of Forest Fragments Project' (now Smithsonian Institute) provided free accommodation and the use of computer facilities in Manaus.

Marlene Freitas da Silva and William Rodrigues offered the full support of INPA Botânica, Manaus, for the identification and export of the botanical material, as did the UTAM institute. The comprehensive botanical knowledge of INPA's 'tecnicos' Luiz and Dionizio Coelho and José Ramos, and their enthusiastic assistance, enabled many sterile specimens to be determined which might otherwise have remained unidentified. The following taxonomic specialists assisted with identification: D.C. Daly (Burseraceae), A. Gentry (Bignoniaceae-part), A.J.M. Leeuwenberg (Apocynaceae — part), G.P. Lewis (Leguminosae —part), P.J.M. Maas (Annonaceae — part), A.M.W. Mennega (Hippocrateaceae), S.A. Mori (Lecythidaceae), T.D. Pennington, (Meliaceae, Sapotaceae, Sapindaceae & *Inga* — part), G.T. Prance (Chrysobalanaceae), W.A. Rodrigues (Myristicaceae), C. Stace (Combretaceae — part), H. van der Werff (Lauraceae) and J.J. Wurdack (Melastomataceae). S.G.Baines, W. Balée, B.M. Boom, P. Grenand, J.M. Lock, D.A. Posey, G.T. Prance, J.A. Ratter and R.E. Schultes generously gave their time to examine and comment upon the manuscript. The phytosociological data (used to identify the dominant families in the hectare plot) were processed with the FITOPAC computer programme, developed and kindly lent by George Shepherd (University of Campinas).

The preparation of this paper was funded by the Bentham-Moxon Trust, Royal Botanic Gardens, Kew. Pierre Grenand kindly authorized the use of some of his unpublished Waimiri Atroari data, and contributed much useful advice. Dr J.A. Ratter and the Royal Botanic Garden, Edinburgh offered valuable assistance and facilities during the final stages of its preparation.

LITERATURE CITED

1 Acero Duarte, L.E. (1979). Principales plantas útiles de la Amazonia Colombiana. Proyecto Radargramétrico del Amazónas, Bogotá, Colombia.

2 Acero Duarte, L.E. (1982). Propiedades, usos y nominación de especies vegetales de la Amazonia Colombiana. Corporacion Araracuana, DAINCO, Bogotá.

3 Acevedo-Rodriguez, P. (1990). The occurrence of piscicides and stupefactants in the plant kingdom. Adv. in Econ. Bot. 8: 1–23.

4 Aguiar, J.P.C., Marinha, H.A., Rebelo, Y.S. & Shrimpton, R. (1980). Aspectos nutritivos de alguns frutas da Amazônia. Acta Amazonica 10 (4): 755–758.

5 Agurell, S., Holmstedt, B.O., Lindgren, J-E. & Schultes, R.E. (1969). Alkaloids in species of *Virola.* Acta Chem. Scand. 23: 903–916.

6 Allorge, L. (1985). Monographie des Apocynacées Tabernaemontanoidées Americaines. Mem. Mus. Natn. Hist. nat. Paris, n.s. 30: 1–216.

7 Altman, R.F.A. (1958). A presença de escatol no tachy. INPA Publ. (Quimica) No. 2.

8 Altschul, S. von R. (1973). Drugs and foods from little-known plants. Harvard University Press, Cambridge, USA.

9 Altschul, S. von R. & Lipp, F. (1982). New sources for drugs and foods from the New York Botanical Garden herbaria. Harvard University Press, Cambridge, USA.

10 Amorozo, M.C. de M. & Gély, A. (1988). Uso de plantas medicinais por caboclos do baixo Amazonas. Bol. Mus. Pará. Emílio Goeldi, Sér. Bot. 4 (1): 47–131.

11 Amshoff, G.J.H. (1939). Papilionaceae. Flora of Suriname (ed. A. Pulle) 2 (2): 1–257. J.H. de Bussy, Amsterdam.

12 Anderson, A.B. (1978). The names and uses of palms among a tribe of Yanomama Indians. Principes 22: 30–41.

13 Anderson, A.B. & Posey, D.A. (1989). Management of a tropical scrub savanna by the Gorotire Kayapó of Brazil. Adv. in Econ. Bot. 7: 159–173.

14 Andersson, L. (1977). The genus *Ischnosiphon* (Marantaceae). Opera Bot. 43: 1–113.

15 Arditti, J. & Rodriguez, E. (1982). *Dieffenbachia*: Uses, abuses and toxic constituents. J. Ethnopharm. 5: 293–302.

16 Arndt, R.R., Brown, S.H., Ling, N.C., Roller, P., Djerassi, C., Ferreira, J.M., Gilbert, B., Miranda, E.C., Flores, S.E., Duarte, A.P. & Carrazzoni, E.P. (1967). The alkaloids of six *Aspidosperma* species. Phytochem. 6: 1653.

17 Auterhoff, H. & Pankow, E. (1968). Inhaltstoffe von Muira puama. Arch. Pharm. (Berlin) 301: 481–489.

18 Ayres, J.M. (1981). Observações sobre a ecologia e o comportamento dos cuxiús. CNPq/INPA/FUA, Manaus.

19 Azevedo Lima, R.R.O. de (1990). Levantamentos preliminares sobre as roças Waimiri Atroari. Unpublished report to Programa Waimiri Atroari, Manaus. 7pp.

20 Baines, S.G. (1988). 'É a FUNAI que sabe': A frente de atração Waimiri Atroari. Unpublished Doctorate Thesis, University of Brasília, Brazil.

21 Baines, S.G. (1991). The Waimiri-Atroari and the Paranapanema company. Critique of Anthropology 11 (2): 143–153.
22 Baker, M., Neill, D., Palacios, W. & Zaruma, J. (1987). Plant resources of Amazonian Ecuador. 2nd Ann. Rep. of Flora Del Ecuador. MBG/NYBG/ USAID/DINAF, Ecuador.
23 Balée, W. (1986). Análise preliminar de inventário florestal e a etnobotânica Ka'apor. Bol. Mus. Pará. Emílio Goeldi, Sér. Bot. 2 (2): 141–167.
24 Balée, W. (1987). A Etnobotânica quantitativa dos Indios Tembé. Bol. Mus. Pará. Emílio Goeldi, Sér. Bot. 3 (1): 29–50.
25 Balée, W. (1988). Indigenous adaptation to Amazonian palm forests. Principes 32 (2): 47–54.
26 Balée, W. (1989a). The culture of Amazonian forests. Adv. in Econ. Bot. 7: 1–21.
27 Balée, W. (1989b). Nomenclatural patterns in Ka'apor ethnobotany. J. Ethnobiol. 9 (1): 1–24.
28 Balée, W. & Daly, D.C. (1990). Resin classification by the Ka'apor Indians. Adv. in Econ. Bot. 8: 24–34.
29 Balée, W. & Gély, A. (1989). Managed forest succession in Amazonia: the Ka'apor case. Adv. in Econ. Bot. 7: 129–158.
30 Balick, M.J. (1979). Economic botany of the Guahibo. I. Palmae. Econ. Bot. 33 (4): 361–376.
31 Balick, M.J. (1986). Systematics and economic botany of the *Oenocarpus-Jessenia* (Palmae) complex. Adv. in Econ. Bot. 3: 1–140.
32 Balick, M.J. (1988). The use of palms by the Apinaye and Guajajara Indians of northeastern Brazil. Adv. in Econ. Bot. 6: 65–90.
33 Balick, M.J. & Gershoff, S.N. (1981). Nutritional evaluation of the *Jessenia bataua* palm: source of high quality protein and oil from tropical America. Econ. Bot. 35 (3): 261–271.
34 Barbosa Rodrigues, J. (1885). Rio Jauapery — Pacificação dos Crichanás. Imprensa Nacional, Rio de Janeiro, Brazil.
35 Berg, C.C. (1972). Olmedieae, Brosimeae (Moraceae). Fl. Neotrop. Monogr. 7: 1–228.
36 Berg, C.C. & Akkermans, R.W.A.P. (1985). Studies on the flora of the Guianas, 14. New taxa and combinations in *Sorocea* (Moraceae) and a key to its species. Proc. Kon. Ned. Akad. Wet. 88 (4): 381–394.
37 Berg, C.C., Vázquez Avila, M. & Kooy, F. (1984). *Ficus* species of Brazilian Amazonia and the Guianas. Acta Amazonica 14 (1/2): 159–194.
38 Berg, C.C., Akkermans, R.W.A.P. & Heusden, E.C.H. van (1990). Cecropiaceae: *Coussapoa* and *Pourouma*, with an introduction to the family. Fl. Neotrop. Monogr. 51: 1–208.
39 Berg, M.E. van den (1982). Plantas medicinais na Amazônia. Contribuição ao seu conhecimento sistemático. CNPQ/PTU, Belém, Brazil.
40 Berg, M.E. van den & Silva, M.L.H. da (1984). Plantas medicinais do Amazonas. Anais, I Simpósio do Tropico Úmido: 127–133.
41 Berg, M.E. van den & Silva, M.L.H. da (1988). Contribuição ao conhecimento da flora medicinal de Roraima. Acta Amazonica 18 (1–2) Supl.: 23–35.

42 Berlin, B. (1976). The concept of rank in ethnobiological classification: some evidence from Aguaruna folk botany. American Ethnologist 3 (3): 381–399.
43 Berlin, B., Breedlove, D.E. & Raven, P.H. (1974). Principles of Tzeltal plant classification. Academic Press, New York, USA.
44 Bézanger-Beauquesne, L., Pinkas, M. & Torck, M. (1986). Les plantes dans la therapeutique moderne. Editeur Maloine, France.
45 Bisset, N.G. (1989). Arrow and dart poisons. J. Ethnopharm. 25 (1): 1–41.
46 Bisset, N.G. & Hylands, P.J. (1977). Cardiotonic glycosides from the latex of *Naucleopsis mello-barretoi*, a dart-poison plant from north-west Brazil. Econ. Bot. 31: 307–311.
47 Bodmer, R.E. (1990). Responses of ungulates to seasonal inundations in the Amazon floodplain. J. Trop. Ecol. 6: 191–201.
48 Boom, B.M. (1987). Ethnobotany of the Chácobo Indians, Beni, Bolivia. Adv. in Econ. Bot. 4: 1–68.
49 Boom, B. (1990). Useful plants of the Panare Indians of Venezuelan Guayana. Adv. in Econ. Bot. 8: 57–76.
50 Branch, L.C. & Silva, M.F. da (1983). Folk medicine of Alter do Chão, Pará, Brazil. Acta Amazonica 13 (5–6): 737–797.
51 Braz Filho, R., Gottlieb, O.R., Morães, A.A. de, Pedreira, G. & Pinho, L.V. (1977). The chemistry of Brazilian Myristicaceae IX. Isoflavonoids from Amazonian species. Lloydia 40: 236.
52 Brewer-Carias, C. & Steyermark, J.A. (1976). Hallucinogenic snuff drugs of the Yanomamo Caburiwe-Teri in the Cauaburi River, Brazil. Econ. Bot. 30 (1): 57–66.
53 Brouk, B. (1975). Plants consumed by man. Academic Press, London.
54 Brücher, H. (1979). Useful plants of neotropical origin, and their wild relatives. Springer-Verlag, Berlin.
55 Buckley, J.P., Theobald, R.J. & Cravero, I. (1973). Preliminary pharmacological evaluation of extracts of Takini: *Helicostylis tomentosa* and *H. pedunculata*. Lloydia 36 (3): 341–345.
56 Buschbacher, R.J. (1986). Tropical deforestation and pasture development. Bioscience 36 (1): 22–28.
57 Campbell, D.G., Daly, D.C., Prance, G.T. & Maciel, U.N. (1986). Quantitative ecological inventory of terra firme and varzea tropical forest on the Rio Xingu, Brazilian Amazon. Brittonia 38: 369–393.
58 Carbajal, D., Casaco, A., Arruzazabala, L., Gonzalez, R. & Fuentes, V. (1991). Pharmacological screening of plant decoctions commonly used in Cuban folk medicine. J. Ethnopharm. 33: 21–24.
59 Carneiro, R.L. (1978). The knowledge and use of rain forest trees by the Kuikuru Indians of Central Brazil. In The Nature and Status of Ethnobotany (ed. R. Ford), pp. 201–216. University of Michigan Press, Michigan, USA.
60 Carvalho, J.F.F. de. (1982). Waimiri Atroari, a história que ainda não foi contada. Brasília, Brazil.
61 Cavalcante, P.B. (1978). Contribuição ao conhecimento das Gnétaceas de Amazônia (Gimnospermas). Acta Amazonica 8 (2): 201–215.
62 Cavalcante, P.B. (1988). Frutas comestiveis do Amazônia. MPEG, Belém, Brazil.

63 Cavalcante, P.B. & Frikel, P. (1973). A farmacopéia Tiriyó. MPEG, Belém, Brazil.
64 Chai, P.P.K. (1978). Ethnobotany. Part II. Sarawak Museum Journal 26 (47): 243–270.
65 Chernela, J.M. (1987). Os cultivares de mandioca na área do Uaupés (Tukâno). In Suma Etnológica Brasileira (Coord. B. Ribeiro) Vol. 1 (Etnobiologia), pp. 151–158. Vozes/FINEP, Petrópolis, Brazil.
66 Chu, E.P. & Figueiredo-Ribeiro, R.C.L. (1991). Native and exotic species of *Dioscorea* in Brazil. Econ. Bot. 45 (4): 467–479.
67 Clayton, W.D. & Renvoize, S.A. (1986). Genera Graminum - Grasses of the world. Royal Botanic Gardens, Kew.
68 Clement, C.R. (1988). Domestication of the pejibaye palm (*Bactris gasipaes*): past and present. Adv. in Econ. Bot. 6: 155–174.
69 Cocco, P.L. (1972). Ijewei-Teri, quince anos entre los Yanomamos. Escuela Tecnica Popular Dr Bosco Boleita, Caracas, Venezuela.
70 Colchester, M. (1989). Indian Development in Amazonia: Risks and Strategies. The Ecologist 19 (6): 249–245.
71 Contrato 039/82. (1982). Contrato No. 039/82, que entre si celebram a FUNAI e a empresa Timbó Industria de Mineração Ltda, objetivando a construcão de uma estrada secundaria. FUNAI, Manaus.
72 Cooper, J.M. (1987). Estimulantes e narcóticos. In Suma Etnológica Brasileira (coord. B. Ribeiro) Vol. 1 (Etnobiologia), pp. 101–118. Vozes/FINEP, Petrópolis, Brazil.
73 Corrêa, M.P. (1919). Flora do Brasil. Algumas plantas úteis, suas applicações e distribuição geográfica. Directoria Geral de Estatística, Rio de Janeiro.
74 Corrêa, M.P. (1926–1975). Dicionário das plantas úteis do Brasil e das exóticas cultivadas. Ministério de Agricultura, Rio de Janeiro, Brazil.
75 Cowan, R.S. (1953). A taxonomic revision of the genus *Macrolobium* (Leg. — Caesalpinioideae). Mem. N.Y. Bot. Gard. 8 (4): 257–342.
76 Cowan, R.S. (1967). *Swartzia* (Leguminosae, Caesalpinioideae, Swartzieae). Fl. Neotrop. Monogr. 1: 1–228.
77 Cuatrecasas, J. (1961). A taxonomic revision of the Humiriaceae. Contrib. U.S. Nat. Herb. 35 (2): 25–214.
78 Cuatrecasas, J. (1964). A taxonomic revision of the genus *Theobroma*. Contrib. U.S. Nat. Herb. 35 (6): 379–614.
79 Daly, D.C. (1987). A taxonomic revision of *Protium* (Burseraceae) in eastern Amazonia and the Guianas. Unpublished Ph.D. Dissertation, City University of New York, USA. 469pp.
80 Davis, E.W. & Yost, J.A. (1983a). The ethnomedicine of the Waorani of Amazonian Ecuador. J. Ethnopharm. 9: 273–297.
81 Davis, E.W. & Yost, J.A. (1983b). The ethnobotany of the Waorani of eastern Ecuador. Bot. Mus. Leafl., Harv. Univ. 29 (3): 159–217.
82 Davis, S.H. (1977). Victims of the Miracle: Development and the Indians of Brazil. Cambridge University Press.
83 Denevan, W.M. & Treacy, J.M. (1987). Young managed fallows at Brillo Nuevo. Adv. in Econ. Bot. 5: 8–46.
84 Denevan, W.M., Treacy, J.M., Alcorn, J.R., Padoch, C. & Denslow, J. (1984). Indigenous agroforestry in the Amazon: Bora Indian management of swidden fallows. Interciência 9 (6): 358–365.

85 Ducke, A. (1935). Revision of the genus *Hevea* Aubl., mainly the Brazilian species. Arch. Inst. Biol. Veget. Rio de Janeiro 2 (2): 217–246.
86 Dufour, D.L. (1988). Cyanide content of cassava (*Manihot esculenta*, Euphorbiaceae) cultivars used by Tukanoan Indians of northwest Amazonia. Econ. Bot. 42 (2): 255–266.
87 Duke, J.A. (1986). Isthmian ethnobotanical dictionary. Scientific Publishers, Jodhpur, India.
88 Earle Smith, C. (1954). The New World species of *Sloanea* (Elaeocarpaceae). Contrib. Gray Herb. 175: 1–114.
89 Elisabetsky, E. (1987). Etnofarmacologia de algumas tribos brasileiras. In Suma Etnológica Brasileira (coord. B. Ribeiro) Vol. 1 (Etnobiologia), pp. 135–148. Vozes/FINEP, Petrópolis, Brazil.
90 Elisabetsky, E. & Posey, D.A. (1989). Use of contraceptive and related plants by the Kayapó Indians. J. Ethnopharm. 26: 299–316.
91 Encarnación, F. & Spichiger, R. (1982). Las Bombacáceas del arborétum Jenaro Herrera (provincia de Requena, departamento de Loreto, Perú). Candollea 37 (1): 1–15.
92 Engler, A. *et al.* (1900–). Das Pflanzenreich. Regni Vegetabilis Conspectus. Wilhelm Engelmann, Leipzig.
93 Ernst, A. (1888). Enumeración de las plantas empleadas en los diferentes paises para el método de pesca conocido bajo el nombre de 'embarbascar'. Rev. Cient. Univ. Centr. Venezuela 1 (9): 318–327.
94 Exell, A.W. & Stace, C.A. (1963). A revision of the genera *Buchenavia* and *Ramatuela*. Bull. Brit. Mus. (Nat. Hist.) Botany 3 (1): 1–46.
95 Eyma, P.J. (1941). Guttiferae. Flora of Suriname (ed. A. Pulle) 3 (1): 65–118. J.H. de Bussy, Amsterdam.
96 Fanshawe, D.B. (1948a). Forestry products of British Guiana. Part 1. Principal timbers. Forestry Bull. 1: 1–101.
97 Fanshawe, D.B. (1948b). Forestry products of British Guiana. Part 2. Minor forest products. Forestry Bull. 2: 1–81.
98 Fearnside, P.M. (1989). Forest management in Amazonia: the need for new criteria in evaluating development options. Forest Ecology and Management 27: 61–79.
99 Fearnside, P.M. (1990). Balbina. Lições trágicas na Amazônia. Ciência Hoje 11 (64): 37–42.
100 Fidalgo, O. (1965). Conhecimento micológico dos Índios Brasileiros. Rickia 2: 1–10.
101 Fidalgo, O. & Prance, G.T. (1976). The ethnomycology of the Sanama Indians. Mycologia 68 (1): 201–210.
102 Flores Paitan, S. (1987). Old managed fallows at Brillo Nuevo. Adv. in Econ. Bot. 5: 53–66.
103 Foldats, E. (1970). Orchidaceae. Flora de Venezuela 15 (5): 9–598. Instituto Botánico, Caracas.
104 Forero Pinto, L.E. (1980). Etnobotánica de las comunidades indígenas Cuna y Wauana, Chocó (Colombia). Cespedesia 9 (33–34): 115–325.
105 Fries, R.E. (1939). Revision der arten einiger Anonaceen-gattungen. Acta Hort. Berg. 12: 1–577.
106 FUNAI (1975). A morte da sertanista Gilberto. Informativo FUNAI 13: 3–9.

107 García-Barriga, H. (1974). Flora medicinal de Colombia. Vol. 1. Instituto de Ciéncias Naturales, Bogotá.

108 Garzon, N.L., Cuca, S.L.E., Yoshida, M. & Gottlieb, O.R. (1987). Flavonolignoid from the fruit of *Iryanthera laevis*. Phytochem. 26 (10): 2835–2837.

109 Gentry, A.H. (1977). Flora of Ecuador. 178: Bignoniaceae. Opera Bot. (Ser. B) 7: 1–173.

110 Glassman, S.F. (1972). A revision of B.E. Dahlgren's index of American palms. Phanerogamarum Monographiae Tomus VI, Verlag J. Cramer, Lehre, Germany.

111 Glenboski, L.L. (1983). The ethnobotany of the Tukuna Indians, Amazónas, Colombia. Biblioteca José Jerónimo Triana No. 4, Universidad Nacional de Colombia, Bogotá.

112 Godoy, R.L. de O., Lima, P.D. de D.D., Pinto, A.C. & Aquino Neto, F.R. de (1989). Diterpenoids from *Dipteryx odorata*. Phytochem. 28 (2): 642–644.

113 Gonçalves de Lima, O. (1975). Pulque, balché e pajauaru na etnobiologia das bebidas e dos alimentos fermentados. Universidade Federal de Pernambuco, Brazil.

114 Gottlieb, O. (1979). Chemical studies on medicinal Myristicaceae. J. Ethnopharm. 1 (4): 309–323.

115 Granja e Barros, M.A. (1982). Flora medicinal do Distrito Federal. Bras. Flores. 12 (50): 35–44.

116 Gray, A. (1990). Indigenous peoples and the marketing of the rainforest. The Ecologist 20 (6): 223–227.

117 Graziano, M.M., Widmer, G.A. & Coussio, J.D. (1967). Isolation of tyramine from five Argentine species of Loranthaceae. Lloydia 30 (3): 242–244.

118 Greenbaum, L. (1989). Plundering the timber on Brazilian Indian reservations. Cultural Survival Quarterly 13 (1): 23–26.

119 Greenway, P.J., Wallace, G.B., Wallace, M.M. & Khomo, E.V.R. (1953). The Pawpaw. Its botany, cultivation, diseases and chemistry. Tanganyika Department of Agriculture, Dar Es Salaam

120 Grenand, P. (1980). Introduction à l'étude de l'univers Wayãpi. Ethnoécologie des Indiens du Haut-Oyapock (Guyane Française). Langues et Civilisations a Tradition Orale 40. SELAF, Paris.

121 Grenand, F. & Ferreira, E.H. (1989). Pequeno dicionário da lingua geral. Seduc Série Amazonas; Cultura Regional 6. Manaus, Brazil.

122 Grenand, P., Moretti, C. & Jacquemin, H. (1987). Pharmacopées traditionelles en Guyane. ORSTOM, Paris, France.

123 Gribel, R. (1990). The Balbina disaster: the need to ask why? The Ecologist 20 (4): 133–135.

124 Grimé, W.E. (1979). Ethno-botany of the black Americans. Reference Publications, Michigan, USA.

125 Gross, D.R., Eiten, G., Flowers, N.M., Leoi, F.M., Ritter, M.L. & Werner, D.W. (1979). Ecology and acculturation among native peoples of central Brazil. Science 206: 1043–1050.

126 Hartmann, T. (1967). Nomenclatura botânica dos Borôro. USP, São Paulo, Brazil.

127 Heath, E.G. & Chiara, V. (1977). Brazilian Indian archery. Simon Archery Foundation, Manchester.

128 Heckel, E. (1897). Les plantes médicinales et toxiques de la Guyane Française. Protat Frères, Maçon, France.
129 Heim, F. (1903). Etudes sur la flore économique et les produits végétaux de la Guyane Française. Comité de la Guyane Française, Paris, France.
130 Hekking, W.H.A. (1988). Violaceae Part 1: *Rinorea* and *Rinoreocarpus*. Fl. Neotrop. Monogr. 46: 1–208.
131 Hemming, J. (1987). Amazon frontier: The defeat of the Brazilian Indians. MacMillan, London.
132 Henderson, R.E. (1990). Arecaceae. Part 1. Introduction and the Iriarteinae. Fl. Neotrop. Monogr. 53: 1–101.
133 Henry, T.A. (1939). The plant alkaloids. J. & A. Churchill, London.
134 Hildyard, N. (1989). Adios Amazonia? A report from the Altamira gathering. The Ecologist 19 (2): 53–62.
135 Hocquemiller, R., Cavé, A., Jacquemin, H., Touche, A. & Forgacs, P. (1982). Alcaloides des Anonacées. XXXVI: alcaloides de l'*Annona crassiflora* Mart. Plant. Med. Phytother. 16 (1): 4–6.
136 Hodge, W.H. & Taylor, D. (1957). The ethnobotany of the Island Caribs of Dominica. Webbia 12 (2): 513–644.
137 Holland, J.H. (1937). Overseas plant products. John Bale, Sons & Curnow, London.
138 Holmstedt, B., Lindgren, J.E., Plowman, T., Schultes, R.E. & Tovar, O. (1980). Indole alkaloids in Amazonian Myristicaceae: field and laboratory research. Bot. Mus. Leafl., Harv. Univ. 28 (3): 215–234.
139 Hopkins, H.C.F. (1986). *Parkia* (Leguminosae: Mimosoideae). Fl. Neotrop. Monogr. 43: 1–124.
140 Howes, F.N. (1930). Fish-poison plants. Kew Bull. 4: 129–153.
141 Hunter, R.J. (1991). Observations on the growth, ecology and uses of *Minquartia guianensis*, a humid tropical tree. Int. Tree Crops J. 6: 221–238.
142 Im Thurn, E.F. (1883). Among the Indians of Guiana. Kegan Paul, London.
143 Irvine, D. (1989). Succession management and resource distribution in an Amazonian rainforest. Adv. in Econ. Bot. 7: 223–237.
144 Izawa, K. (1975). Foods and feeding behaviour in monkeys in the Upper Amazon basin. Primates 16 (3): 295–316.
145 Izawa, K. (1979). Foods and feeding behaviour of wild black-capped capuchin (*Cebus apella*). Primates 20 (1): 57–76.
146 Jansen-Jacobs, M.J. (1986). Dilleniaceae. Flora of Suriname 3 (1–2) (eds. A.L. Stoffers & J.C. Lindman): 475–484. E.J. Brill, Leiden.
147 Janzen. D.H. (1986). The future of tropical ecology. Ann. Rev. Ecol. Syst. 17: 305–324.
148 Jossang, A., Dubaele, B., Cav, A., Bartoli, M-H. & Briel, H. (1991). Annomontacine: une nouvelle acétogénine τ-lactone-monotetrahydrofurannique cytotoxique de l'*Annona montana*. Lloydia 54 (4): 967–971.
149 Kerr, W.E. (1987). Agricultura e seleções genéticas de plantas. In Suma Etnológica Brasileira (coord. B. Ribeiro) Vol. 1 (Etnobiologia), pp. 159-171. Vozes/FINEP, Petrópolis, Brazil.
150 Killip, E.P. (1938). The American species of Passifloraceae. Fieldiana Bot. 19 (2): 1–613.

151 Killip, E.P. & Smith, A.C. (1935). Some American plants used as fish poisons. Bureau of Plant Industry, U.S. Department of Agriculture, USA.

152 Kleinhoonte, A. (1940). Mimosaceae. Flora of Suriname (ed. A. Pulle) 2 (2): 258–331. J.H. de Bussy, Amsterdam.

153 Krukoff, B.A. (1972). American species of *Strychnos*. Lloydia 35 (3): 193–271.

154 Krukoff, B.A. & Smith, A.C. (1937). Rotenone-yielding plants of South America. Amer. J. Bot. 24 (9): 573–587.

155 Kubitzki, K. (1970). Die gattung *Tetracera* (Dilleniaceae). Mitt. Bot. München 8: 1–98.

156 Kubitzki, K. & Brenner, S. (1982). Lauraceae I (*Aniba* and *Aiouea*). Fl. Neotrop. Monogr. 31: 1–125.

157 Kvist, L.P. & Holm-Nielsen, L.B. (1987). Ethnobotanical aspects of lowland Ecuador. Opera Bot. 92: 83–107.

158 Lago, R.C.A., Pereira, D.A., Siqueira, F.A.R., Szpiz, R.R. & Oliveira, J.P. de (1986/7). Estudo preliminar das sementes e do oleo de cinco espécies da Amazônia. Acta Amazonica 16/17 (1): 369–376.

159 Lanjouw, J. (1936). Studies in Moraceae II. The genus *Clarisia* Ruiz et Pavón and its synonyms, with a discussion of the generic name. Rec. Trav. Bot. Neerl. 33: 254–276.

160 Lanjouw, J. & Van Heerdt, P.F. (1941a). Marcgraviaceae. Flora of Suriname (ed. A. Pulle) 3 (1): 373–385. J.H. de Bussy, Amsterdam.

161 Lanjouw, J. & Van Heerdt, P.F. (1941b). Quiinaceae. Flora of Suriname (ed. A. Pulle) 3 (1): 355–365. J.H. de Bussy, Amsterdam.

162 La Rotta, C. (1983). Observaciones etnobotánicas sobre algunas especies utilizadas por la comunidad indígena Andoque (Amazónas, Colombia). DAINCO, Bogotá, Colombia.

163 LeCointe, P. (1934). Arvores e plantas úteis da Amazônia Brasileira. Livraria Classica, Belém.

164 Lewis, W.H., Elvin-Lewis, M. & Gnerre, M.C. (1987). Introduction to the ethnobotanical pharmacopoeia of the Amazonian Jívaro of Peru. In Medicinal and poisonous plants of the tropics (comp. A.J. Leeuwenberg), pp. 96–103. Pudoc, Wageningen.

165 Loureiro, A.A. & Lisboa, P.L.B. (1979). Madeiras do Município de Aripuana e suas utilidades (Mato Grosso). Acta Amazonica 9 (1) Supl.: 1–88.

166 Loureiro, A.A. & Silva, M.F. da (1968). Catálogo das madeiras da Amazônia (2 Vols). SUDAM, Belém.

167 Lowie, R.H. (1948). The Tropical forest tribes. In: The handbook of South American Indians (ed. J.H. Steward) Vol. 3, Smithsonian Institute, Washington. pp. 1–56.

168 Luz, A.I.R., Zoghbi, M.G.B., Ramos, L.S., Maia, J.G.S. & Silva, M.L. (1984). Essential oils of some Amazonian Zingiberaceae, 3. Genera *Alpinia* and *Renealmia*. Lloydia 47: 907–908.

169 Maas, P.J.M. (1977). *Renealmia* (Zingiberaceae - Zingiberoideae). Fl. Neotrop. Mongr. 18: 1–218.

170 Mabberley, D.J. (1987). The plant book. Cambridge University Press.

171 Maclean, C. (1977). Island on the edge of the world. Revised edition. Canongate, Edinburgh.

172 MacRae, W.D. & Towers, G.H.N. (1984). An ethnopharmacological examination of *Virola elongata* bark. J. Ethnopharm. 12 (1): 75–92.
173 MacRae, W.D. & Towers, G.H.N. (1985). Non-alkaloidal constituents of *Virola elongata* bark. Phytochem. 24 (3): 561–566.
174 McVaugh, R. (1969). The botany of the Guayana Highland — Part VIII. Myrtaceae. Mem. N.Y. Bot. Gard. 18 (2): 55–286.
175 MAREWA (1983). Resistência Waimiri Atroari. Itacoatiara, Brazil.
176 Markgraf, F. von (1941). Die gattung *Lacmellea* H. Karsten. Notizblatt. Bot. Gard. Berlin 15: 615–629.
177 Marles, R.J. & Farnsworth, N.R. (1989). Isolation of a novel cytotoxic polyacetylene from a traditional antihelmintic plant, *Minquartia guianensis*. Lloydia 52 (2): 261–266.
178 Mayo, S.J. (1991). A revision of *Philodendron* subgenus *Meconostigma* (Araceae). Kew Bull. 46 (4): 601–681.
179 Miller, R.P., Wandelli, E.V. & Grenand, P. (1989). Conhecimento e utilização da floresta pelos Índios Waimiri Atroari do Rio Camanaú -Amazonas (I). Acta Bot. bras. 3(2) Supl.: 47–56.
180 Milton, K. (1980). The foraging strategy of howler monkeys. A study in primate economics. Colombia University Press, New York.
181 Mittermeier, R.A. & Roosmalen, M.G.M. van (1981). Preliminary observations on habitat utilization and diet in eight Surinam monkeys. Folia Primatol. 36 (1–2): 1–39.
182 Monachino, J. (1943). A revision of *Couma* and *Parahancornia*. Lloydia 6 (4): 229–247.
183 Monachino, J. (1949). A revision of *Ryania* (Flacourtiaceae). Lloydia 12: 1–29.
184 Moretti, C. & Grenand, P. (1982). Les nivrées ou plantes ichtyotoxiques de la Guyane Française. J. Ethnopharm. 6 (2): 139–160.
185 Mori, S.A., Prance, G.T. & Zeeuw, C.H. de (1990). Lecythidaceae - Part II. The zygomorphic-flowered New World genera. Fl. Neotrop. Monogr. 21 (II): 1–376.
186 Mors, W.B. & Rizzini, C.T. (1966). Useful plants of Brazil. Holden-Day, San Francisco, USA.
187 Morton, J.F. (1981). Atlas of medicinal plants of Middle America. Charles Thomas, Springfield, USA.
188 Moskovits, D.K. (1985). The behaviour and ecology of two Amazonian tortoises, *Geochelone carbonaria* and *G. denticulata* in north-western Brasil. Unpublished doctoral thesis, Div. of Biology, University of Chicago, USA. 328 pp.
189 Nakano, T. & Suarez, M. (1970). Studies on the neutral constituents of the bark of *Dipteryx odorata*. Planta Med. 18 (1): 79–81.
190 Nye, M.M. (1991). The mis-measure of manioc (*Manihot esculenta*, Euphorbiaceae). Econ. Bot. 45 (1): 47–57.
191 Ouensanga, C. (1983). Plantes médicinales et remèdes Créoles. Vol. I. Emile Desormeaux, Paris, France.
192 Parodi, J.L. (1988). The use of palms and other native plants in non-conventional, low cost rural housing in the Peruvian Amazon. Adv. in Econ. Bot. 6: 119–129.
193 Peixoto, A.R. (1973). Plantas oleaginosas arbóreas. Livraria Nobel, São Paulo, Brazil.

194 Pellicciari, R., Ardon, A. & Bellavita, V. (1972). Triterpenes from *Pouteria caimito*. Planta Med. 22 (2): 196–200.
195 Pennington, T.D. (1990). Sapotaceae. Fl. Neotrop. Monogr. 52: 1–770.
196 Pennington, T.D., Styles, B.T. & Taylor, D.A.H. (1981). Meliceae. Fl. Neotrop. Monogr. 28: 1–470.
197 Pereira, V.S. (1990). Mineradora provoca intrigas para expulsar pesquisador. Boletim da ABA 9: 12–13.
198 Perez Arbelaez, E. (1947). Plantas utiles de Colombia. Contraloria General de la Republica, Bogotá.
199 Perkins, J.R. (1901). Beiträge zur kenntnig der Monimiaceae. III. Monogrpahie der gattung *Siparuna*. Bot. Jahrb. (ed. A. Engler) 28: 660–705. Wilhelm Engelmann, Leipzig.
200 Pernia, N.E. de (1982). Cristales y silice en maderas dicotiledoneas de Latinoamerica. Pittieria 11: 13–65.
201 Pesce, C. (1985). Oil palms and other oilseeds of the Amazon. Reference Publications, Michigan, USA.
202 Peters, C.M. (1985). Annotated list of the native fruit trees of the Lower Ucayali. Unpublished Progress Report. IEB/NYBG, New York, USA. 23pp.
203 Peters, C.M., Gentry, A.H. & Mendelsohn, R.O. (1989). Valuation of an Amazonian forest. Nature 339: 655–656.
204 Pinedo-Vasquez, M., Zarin, D., Jipp, P. & Chota-Inuma, J. (1991). Use-values of tree species in a communal forest reserve in northeast Peru. Biological Conservation 4 (4): 405–416.
205 Pinheiro, C.U.B. & Balick, M.J. (1987). Brazilian Palms. Contr. New York Bot. Gard. 17: 1–50.
206 Planchon, L. (1888). Étude sur les produits de la famille des Sapotées. Impr. Centrale du Midi, Montpellier, France.
207 Plotkin, M.J. & Balick, M.J. (1984). Medicinal uses of South American palms. J. Ethnopharm. 10 (2): 157–179.
208 Plotkin, M.J., Boom, B.M. & Allison, M. (1991). The ethnobotany of Aublet's Histoire des plantes de la Guiane Française (1775). Monographs in Systematic Botany 35.
209 Porantim (1990). Denúncia: Paranapanema, mina de corrupção. Porantim (April 1990): 7.
210 Posey, D.A. (1983). Indigenous knowledge and development: an ideological bridge to the future. Ciênc. e Cult. (S. Paulo) 35 (7): 877–894.
211 Posey, D.A. (1984). A preliminary report on diversified management of tropical forest by the Kayapó Indians of the Brazilian Amazon. Adv. in Econ. Bot. 1: 112–126.
212 Posey, D.A. (1987). Etnobiologia: teoria e prática. In Suma Etnológica Brasileira (coord. B. Ribeiro) Vol. 1 (Etnobiologia), pp. 15–25. Vozes/FINEP, Petrópolis, Brazil.
213 Posey, D.A. (1990a). The application of ethnobiology in the conservation of dwindling natural resources: lost knowledge or options for the survival of the planet. In Ethnobiology: implications and applications (ed. D.A. Posey & W.L. Overal), pp. 47–60. Belém, Brazil.
214 Posey, D.A. (1990b). Intellectual property rights and just compensation for indigenous knowledge. Anthropology Today 6 (4): 13–16.

215 Prance, G.T. (1972a). An ethnobotanical comparison of four tribes of Amazonian Indians. Acta Amazonica 2 (2): 7–27.
216 Prance, G.T. (1972b). Chrysobalanceae. Fl. Neotrop. Monogr. 9: 1–410.
217 Prance, G.T. (1984). The use of edible fungi by Amazonian Indians. Adv. in Econ. Bot. 1: 127–139.
218 Prance, G.T. (1987). Etnobotânica de algumas tribos Amazônicas. In Suma Etnológica Brasileira (coord. B. Ribeiro) Vol. 1 (Etnobiologia), pp. 119–133. Vozes/FINEP, Petrópolis, Brazil.
219 Prance, G.T. & Mori, S.A. (1979). Lecythidaceae - Part I. The actinomorphic New World Lecythidaceae. Fl. Neotrop. Monogr. 21: 1–270.
220 Prance, G.T. & Mori, S.A. (1983). Dispersal and distribution of Lecythidaceae and Chrysobalanceae. Sonderb. Naturwiss. Ver. Hamburg 7: 163–186.
221 Prance, G.T. & Silva, M.F. da (1973). Caryocaraceae. Fl. Neotrop. Monogr. 12: 1–75.
222 Prance, G.T., Rodrigues, W.A. & Silva, M.F. da (1976). Inventário de um hectare de mata de terra firme km 30 da estrada Manaus-Itacoatiara. Acta Amazonica 6 (1): 9–35.
223 Prance, G.T., Campbell, D.G. & Nelson, B.W. (1977). The ethnobotany of the Paumarí Indians. Econ. Bot. 31 (2): 129–139.
224 Prance, G.T., Balée, W., Boom, B.M. & Carneiro, R.L. (1987). Quantitative ethnobotany and the case for conservation in Amazonia. Conservation Biology 1 (4): 296–310.
225 RADAM (1978). Folha SA 20 Manaus. Levantamento de recursos naturais 18: 1–628. DNPM/RADAMBRASIL, Rio de Janeiro, Brazil.
226 Ratter, J.A. & Furley, P.A. (1983). Properties and uses of native plants. In: Resources and development in Belize (eds G.M. Robinson & P.A. Furley), pp. 157–164. Department of Geography, University of Edinburgh.
227 Ratter, J.A., Pott, A., Vali, J.P., Cátia, N. da C. & Haridasan, M. (1988). Observations on woody vegetation types in the pantanal and at Corumbá, Brazil. Notes RBG Edinb. 45 (3): 503–525.
228 Ribeiro, B.G. & Kenhiri, T. (1989). Rainy seasons and constellations: The Desâna economic calendar. Adv. in Econ. Bot. 7: 97–114.
229 Rico-Arce, M. de L. (1991). New species, combinations and synonyms for *Zygia*, *Cojoba*, *Marmarozylon* and *Pithecellobium* (Leguminosae-Mimosoideae, Ingeae). Kew Bull. 46 (3): 493–521.
230 Rizzini, C.T. (1971). Árvores e madeiras úteis do Brasil. Editora Edgard Blücher, São Paulo, Brazil.
231 Rizzini, C.T. & Mors, W.B. (1976). Botânica econômica Brasileira. USP, São Paulo, Brazil.
232 Rocha, A.I. de, Silva, M.L.H. da, Mourão, A.P. & Cava, M.P. (1968). A presenca de alcaloides em espécies botânicas da Amazônia. CNPQ/INPA, Publicação No. 12, Manaus.
233 Rocha, J. de S., Paula, E.V.C.M. de & Siqueira, M.L. (1988). Flexão estática em amostras pequenas livres de defeitos. Acta Amazonica 18 (1–2): 147–161.
234 Rodrigues, W.A. (1980). Revisão taxonómica das espécies de *Virola* Aublet (Myristicaceae) do Brasil. Acta Amazonica 10 (1) Supl.: 1–127.

235 Rodriguez, E. & Cavin, J.C. (1982). The possible role of Amazonian psychoactive plants in the chemotherapy of parasitic worms - a hypothesis. J. Ethnopharm. 6 (3): 303–309.
236 Rohwer, J.G. von (1986). Prodromus einer monographie der gattung *Ocotea* Aublet (Lauraceae), sensu lato. Mitt. Inst. Allg. Bot. Hamburg 20: 5–278.
237 Romero, J.H.T. (1983). Contribucíon al conocimento de las plantas tintóreas registradas en Colombia. Biblioteca José Jerónimo Triana No. 3, Universidad Nacional de Colombia, Bogotá.
238 Roosmalen, M.G.M. van (1985). Habitat preferences, diet, feeding strategy and social organisation of the black spider monkey in Surinam. Acta Amazonica 15 (3/4): 1–238.
239 Roth, W.E. (1924). An introductory study of the arts, crafts, and customs of the Guiana Indians. Ann. Rep. Bur. Amer. Ethn. 38: 25–745.
240 Rovinski, J.M. & Sneden, A.T. (1984). Furanocoumarins from *Maquira calophylla*. Lloydia 47 (3): 557.
241 Rovinski, J.M., Tewalt, G.L. & Sneden, A.T. (1987). Maquiroside, a new cytotoxic cardiac glycoside from *Maquira calophylla*. Lloydia 50 (2): 211–216.
242 Saleh, N.A.M. & Towers, G.H.N. (1974). Flavonol glycosides of *Norantea guinanesis* flowers. Phytochem. 13: 2012.
243 Salick, J. (1989). Ecological basis of Amuesha agriculture. Adv. in Econ. Bot. 7: 189–212.
244 Sandwith, N.Y. (1956). Contributions to the flora of tropical America: LXII. Spruce's "Bauna" plant from the Amazons. Kew Bull. 1956 (2): 294–296.
245 Schultes, R.E. (1954). A new narcotic snuff from the northwest Amazon. Bot. Mus. Leafl., *Harv. Univ.* 16: 241–260.
246 Schultes, R.E. (1969). De plantis toxicariis e mundo novo tropicale, commentationes V. *Virola* as an orally administered hallucinogen. Bot. Mus. Leafl., Harv. Univ. 22 (6): 229–240.
247 Schultes, R.E. (1974). Palms and religion in the northwest Amazon. Principes 18: 3–21.
248 Schultes, R.E. (1977). Promising structural fiber palms of the Colombian Amazon. Principes 21: 72–82.
249 Schultes, R.E. (1978). De plantis toxicariis e mundo novo tropicale, commentationes XXIII. Notes on biodynamic plants of aboriginal use in the northwestern Amazonia. Bot. Mus. Leafl., Harv. Univ. 26 (5): 177–197.
250 Schultes, R.E. (1979a). De plantis toxicariis e mundo novo tropicale, commentationes XIX. Biodynamic apocynaceous plants of the northwest Amazon. J. Ethnopharm. 1: 165–192.
251 Schultes, R.E. (1979b). Evolution of the identification of the myristicaceous hallucinogens of South America. J. Ethnopharm. 1 (3): 211–239.
252 Schultes, R.E. (1980). De plantis toxicariis e mundo novo tropicale, commentationes XXVI. Ethnopharmacological notes on the flora of northwestern South America. Bot. Mus. Leafl., Harv. Univ. 28 (1): 1–45.
253 Schultes, R.E. (1983a). De plantis toxicariis e mundo novo tropicale, commentationes XXX. Biodynamic guttiferous plants of the northwest Amazon. Bot. Mus. Leafl., Harv. Univ. 29 (1): 49–57.

254 Schultes, R.E. (1983b). De plantis toxicariis e mundo novo tropicale, commentationes XXXII. Notes, primarily on field tests and native nomenclature, on biodynamic plants of the northwest Amazon. Bot. Mus. Leafl., Harv. Univ. 29 (3): 251–272.

255 Schultes, R.E. (1984). De plantis toxicariis e mundo novo tropicale, commentationes XXXIII. Ethnobotanical, floristic and nomenclatural notes on plants of the northwest Amazon. Bot. Mus. Leafl., Harv. Univ. 29 (4): 343–365.

256 Schultes, R.E. & Holmstedt, B. (1971). De plantis toxicariis e mundo novo tropicale, commentationes VIII. Miscellaneous notes on myristicaceous plants of South America. Lloydia 34 (1): 61–78.

257 Schultes, R.E. & Raffauf, R.F. (1990). The healing forest — medicinal and toxic plants of the northwest Amazonia. Historical, ethno- and economic botany series, Vol. 2. Dioscorides Press, Portland, Oregon, USA.

258 Schultes, R.E., Swain, T. & Plowman, T.C. (1977). De plantis toxicariis e mundo novo tropicale, commentationes XVII. *Virola* as an oral hallucinogen among the Boras of Peru. Bot. Mus. Leafl., Harv. Univ. 25 (9): 259–272.

259 Scott, W. (1950). The industrial utilization of sugar cane by-products. Caribbean Commission, Port of Spain, Trinidad.

260 Silva, M.F. da (1991). Waimiri Atroari pararam de crescer. Porantim (December 1991): 6–7.

261 Silva, M.F. da, Lisboa, P.L.B. & Lisboa, R.C.L. (1977). Nomes vulgares de plantas Amazônicas. CNPq/INPA, Manaus, Brazil.

262 Simões, M.F. (1974). Contribuição à arqueologia dos arredores do Baixo Rio Negro, Publ. Avuls. Mus. Pará. Emílio Goeldi 26. Belém, Brazil.

263 Sleumer, H.O. (1984). Olacaceae. Fl. Neotrop. Monogr. 38: 1–159.

264 Smith, A.C. (1940). The American species of Hippocrateaceae. Brittonia 3 (3): 341–555.

265 Smith, A.C. & Wodehouse, R.P. (1937). The American species of Myristicaceae. Brittonia 2 (5): 393–527.

266 Smith, N.J.H. (1980). Anthrosols and human carrying capacity in Amazonia. Ann. Assoc. Amer. Geogr. 70 (4): 553–566.

267 Steyermark, J.A. (1972). The botany of the Guayana Highland — Part IX. Rubiaceae. Mem. N.Y. Bot. Gard. 23: 227–832.

268 Steyermark, J.A. (1974). Rubiaceae. Flora de Venezuela 9 (2): 603–1101. Instituto Botanico, Caracas.

269 Stover, R.H. & Simmonds, N.W. (1987). Bananas. Longman Tropical Agriculture Series. 3rd Edition.

270 Swart, J.J. (1942). A monograph of the genus *Protium* and some allied genera. Rec. Trav. Bot. Neerl. 39: 211–446.

271 Terborgh, J. (1983). Five New World primates. A study in comparative zoology. Princeton University Press, New Jersey, USA.

272 Termo de Compromisso 001/88 (1988). Termo de compromisso no. 001/88 que entre si celebram a empresa Mineração Taboca S.A. e a comunidade indígena Waimiri Atroari objetivando a realização de projetos de desenvolvimento comunitário. FUNAI, Manaus.

273 Termo de Compromisso 001/89 (1989). Termo de compromisso no. 001/89 que entre si celebram a empresa Mineração Taboca S.A. e a comunidade indígena Waimiri Atroari objetivando a realização de projetos de desenvolvimento comunitário. FUNAI, Manaus.
274 Termo de Re-Ratificação 001/88 (1988). Termo de re-ratificaçao no. 001/88 ao Termo de compromisso no. 001/88 que entre si celebram a empresa Mineracão Taboca S.A. e a comunidade indígena Waimiri Atroari objetivando a realização de projetos de desenvolvimento comunitário. FUNAI, Manaus.
275 Tokarnia, C.H., Dobereiner, J. & Silva, M.F. da (1979). Plantas toxicas da Amazônia a bovinos e outros herbívoros. INPA, Manaus, Brazil.
276 Uhe, G. (1974). Medicinal plants of Samoa. A preliminary survey of the use of plants for medicinal purposes in the Samoan Islands. Economic Botany 28 (1): 1–30.
277 Uphof, J.C.T. (1959). Dictionary of economic plants. Hafner, New York.
278 Valle, J.R. & Silva, N.P. (1973). Ichthyotoxicity of cannabinoids. Ciênc. e Cult. (S. Paulo) 25: 647.
279 Vasquez, R. & Gentry, A.H. (1989). Use and misuse of forest-harvested fruits in the Iquitos area. Conservation Biology 3 (4): 350–361.
280 Verheij, E.W.M. & Coronel, R.E. (eds) (1991). Plant resources of South-East Asia. No. 2 — Edible fruits and nuts. Pudoc, Wageningen.
281 Verpoorte, A.U. & Dihal, P.P. (1987). Medicinal plants of Surinam IV. Antimicrobial activity of some medicinal plants. J. Ethnopharm. 21 (3): 315–318.
282 Vickers, W.T. & Plowman, T. (1984). Useful plants of the Siona and Secoya Indians of eastern Ecuador. Fieldiana Bot., n.s. 15: 1–63.
283 Villa Nova, N.A., Salati, E. & Matsui, E. (1976). Estimativa de evapotranspiração na bacia Amazônica. Acta Amazonica 6 (2): 215–228.
284 Wallace, A.R. (1853). Palm trees of the Amazon and their uses. John van Voorst, London.
285 Watt, J.M. & Breyer-Brandwijk, M.G. (1962). The medicinal and poisonous plants of southern and eastern Africa. E. & S. Livingstone, Edinburgh & London.
286 Wessels Boer, J.G. (1988). Palmas indigenas de Venezuela. Pittieria 17: 1–332.
287 Wilbert, W. & Haiek, G. (1991). Phytochemical screening of a Warao pharmacopeia employed to treat gastrointestinal disorders. J. Ethnopharm. 34: 7–11.
288 Woodley, E. (ed.) (1991). Medicinal plants of Papua New Guinea. Part I: Morobe Province. Verlag Josef Margraf, Germany.
289 Woodson, R.E. (1935). Studies in the Apocynaceae IV. The American genera of Echitoideae. Ann. Missouri Bot. Gard. 22 (2): 153–340.
290 Woodson, R.E. (1951). Studies in the Apocynaceae VIII. An interim revision of the genus *Aspidosperma* Mart. & Zucc. Ann. Missouri Bot. Gard. 38: 119–206.
291 Wright, M. (1991). The great rainforest rip-off. Conservation Now (Feb–Mar 1991): 56–57.
292 Wright, W. (1787). An account of the medicinal plants growing in Jamaica. Lond. Med. J. 8 (3): 217–296.

293 Wurdack, J.J. (1980). Melastomataceae. Flora of Ecuador 138. Opera Bot. 13: 1–406.
294 Yost, J.A. & Kelley, P.M. (1983). Shotguns, blowguns, and spears: the analysis of technological efficiency. In: Adaptive responses of native Amazonians (R.B. Hames & W.T. Vickers eds.), pp. 189–224. Academic Press, New York.
295 Yuncker, T.G. (1972). The Piperaceae of Brazil. I - *Piper* - Group I, II, III, IV. Hoehnea 2: 19–366.
296 Zoghbi, M., Das, G.B., Roque, N.F. & Gottlieb, O.R. (1981a). Propacin, a coumarinolignoid from *Protium opacum*. Phytochem. 20 (1): 180.
297 Zoghbi, M., Das, G.B., Roque, N.F. & Gottlieb, O.R. (1981b). Humirianthenolides: new degraded diterpenoids from *Humirianthera rupestris*. Phytochem.20 (7): 1669–1673.

Appendix I

Approximate pronunciation of Waimiri Aroari words and names
(following the orthography used by the education programme of Programa Waimiri Atroari in 1991)

Equivalent Pronunciation

a	as in Portuguese m***a***la
b	as in English ***b***at
d	as in English ***d***og
e	as in French ***ê***tre
h	as in English ***h***at
i	as in Portuguese ***i***lha
j	as in English ***y***et
k	as in English ***k***it
m	as in English ***m***at
n	as in English ***n***ut
o	as in Portuguese far***o***l
p	as in English ***p***ut
r	as in English ***r***at
s	as in English ***s***it
t	as in English ***t***ap
u	as in English y***ou***
w	as in Engliih ***w***et
x	as in English ***sh***e
y	as in French f***eu***
tx	as in English ***ch***urch

Note: In some words, the vowels (including y) are pronounced with a glottal stop, and are written a', e' etc.

Appendix II

Waimiri Atroari use-categories and descriptive 'botanical' terms recorded during the study

aba ieptxi	Medicine for eye infections.
amoty	Useful (bast) fibres in bark.
arawyta wykry[1]	The fruits are eaten by howler monkeys.
are'bri ieptxi	Medicine for the liver.
basa ieptxi	Medicine for sores, boils etc.
dehedehe ieptxi	Medicine for cough.
ia'by ieptxi	Medicine for fevers(?).
iaky ieptxi	Medicine for bite of ponerine ant.
iaweri ieptxi	Medicine for 'madness'.
iaweri ipe	Fruit causes 'madness'.
ietpa iweptxi	Medicine for stomach ache.
ietpa ieptxi	Medicine for pain/fever.
i'ri	Tree bearing edible fruits.
kaduwi ieptxi	Medicine for botfly larval infestation.
kamia wykapry	Used by white men for house-posts.
kamtxa ieptxi	Medicine for localized aches/pains.
ka'na ieptxi	Mosquito medicine.
karawyri kany	Component of healing brew of *marba* ceremony.
keky ieptxi	Medicine for diarrhoea.
kie ieptxi	Medicine for toothache.
kirita ieptxi	?
kory ieptxi	Medicine for 'pus' (septic wounds?).
kyrwy ieptxi	Medicine for snakebite.
kyrwy iepahy ieptxi	Medicine for snakebite.
kwata ieny	Spider monkey sleeping tree.
maba ipa[1]	The fruits are eaten by macaw.
maia iepry	Wood used for large-bladed arrow-heads.
maia pahy ieptxi	Medicine for knife cut.
marba iepra	Plant involved in *marba* ceremony.
maxkwiri ieptxi	Medicine for worms[2].
mei iaty	Firewood for cooking manioc cakes.
mixi kany	Additive for *Maurita*-palm drink.
mydy iepra'ba	Previously used for house construction.
mydy iepry	Used for house construction.
pakia wytypy	Used for killing collared peccaries.
pitxeni kipytxiwy	Medicine for babies (diarrhoea?).

[1] If the fruits of a plant are eaten by a type of monkey, this plant will be described as *arawyta wykry* (*arawyta* = howler monkey, *wyky* = banana, *ry* = possessive suffix), *kwata wykry* etc. If the fruits are eaten by another type of animal, it will be described as *maba ipa* (*maba* = macaw, *ipe* = fruit), *mepri ipa, wiamy ipa* etc.

[2] The word *maxkwiri,* although generally translated as worm, appears (at least in this context) to be also applied to fungal skin infections and (probably) mite infestations.

Appendix II continued

pitxeni kinji'	Medicine for babies.
pyrwa iepry	Wood used for arrow-heads.
pyrwa pyntywy	Glue for arrow-making.
same ieptxi	Medicine for skin infection (white blotches).
sere' ieptxi	Medicine for excessive menstrual bleeding.
swa ieptxi	Medicine for spider bite.
tapany iepy	Hard tree.
wiwe pahy ieptxi	Medicine for 'wood-wound' e.g. stake in foot.
wara' ieny	Tree yielding resin.
warma iepyra	Species involved in preparation of *Ischnosiphon*
wiamy amotyty	Provides fibre for carrying tortoises.
wydi kany	Additive to pottery clay.
wykany	Additive to banana drink.
wyty iaty	Firewood for cooking meat.
xeri ieptxi	Medicine for sting of sting-ray.
xiba imoty	Provides fibre for carrying fish.
yry iepry	Wood used for axe handles.

Appendix III

Scientific and Waimiri Atroari names of animals mentioned in the text.

Agouti	*Dasyprocta agouti*	*akiri*
Anteater	*Myrmecophaga tridactyla*	*sarma*
Botfly	*Dermatobia hominis*	*kaduwi*
Bushmaster	*Lachesis muta*	*kydyty*
Cacique bird	*Cacicus*	*iape'*
Caiman	*Caiman crocodilus*	*iake*
Capuchin monkey	*Cebus apella*	*meky*
Collared peccary	*Tayassu tajacu*	*pakia*
Curassow	*Crax*	*wuky*
Cutiara	*Myoprocta acouchy*	*axwi*
Deer	*Mazama etc.*	*sweri*
Guan	*Penelope*	*mare'*
Howler monkey	*Alouatta seniculus*	*arawyta*
Leafcutter ant	*Atta*	*kymaky*
Macaw	*Ara*	*maba*
Oropendola	*Psarocolius etc.*	*symy*
Paca	*Agouti paca*	*warna*
Pacu	*Characidae*	*paky*
Parakeet	*Pyrrhura etc.*	*xirkiki*
Parrot	*Amazona etc.*	*kyrwaky*
Ponerine ant	*Paraponera*	*iaky*
Saki monkey	*Chiropotes*	*kyxiy*
Scorpion	*Scorpionidae*	*ximri*
Silky anteater	*Cyclopes didactylus*	*iby*
Sloth	*Choleopus/Bradypus*	*wutxiri*
Spider monkey	*Ateles*	*kwata*
Squirrel monkey	*Saimiri sciurius*	*ikamia*
Tapir	*Tapirus terrestris*	*mepri*
Tayra	*Eira barbara*	*wiria*
Tinamou	*Tinamus etc.*	*xuwi/iamy*
Tortoise	*Geochelone*	*wiamy*
Toucan	*Ramphastos*	*kiamky*
Toucanet	*Aulacorhynchus etc.*	*iryryda*
Trumpeter	*Psophia crepitans*	*ikemi*
Tucunare	*Cichla*	*amakra*
Turtle	*Podocnemus*	*warara*
White-lipped peccary	*Tayassu pecari*	*petxi*
Woolly monkey	*Lagothrix lagothricha*	
(monkey)	?	*iarky*
(monkey)	?	*kyxiri*
??		*meiyra*

Appendix IV

Some medicinal plants known to the Waimiri Atroari — arranged according to application. See Table 3 for details.

Wounds (including infected sores)
APOCYNACEAE: *Abuta sandwithiana*. GUTTIFERAE: *Clusia grandiflora*. LORANTHACEAE: Species indet. [GR2761]. MORACEAE: *Ficus guianensis*, *Helianthostylis sprucei*, *Helicostylis tomentosa*, *Naucleopsis amara*, *Naucleopsis mello-barretoi*. MYRISTICACEAE: *Iryanthera laevis*, *Virola theiodora*. PALMAE: *Oenocarpus bacaba*. PIPERACEAE: *Piper consanguineum*, *Piper* sp. [777]. RUBIACEAE: *Palicourea quadrifolia*.

Toothache
ANNONACEAE: *Duguetia cauliflora*. APOCYNACEAE: *Aspidosperma nitidum*. ARACEAE: *Dieffenbachia* sp. [719]. EUPHORBIACEAE *Croton lanjouwensis*. LEGUMINOSAE: *Dipteryx odorata*. MARCGRAVIACEAE: *Marcgravia parviflora*. MYRISTICACEAE: *Virola michelii*. SIMAROUBACEAE: *Picrolemma sprucei*.

Fever/Pain
BIGNONIACEAE: *Callichlamys latifolia*. EUPHORBIACEAE: *Croton lanjouwensis*. GUTTIFERAE: *Clusia* sp. [801]. LEGUMINOSAE: *Parkia nitida*. (MYRISTICACEAE: *Osteophloem platyspermum*?)

Stings and bites (Insects & Arachnids)
ANNONACEAE: *Duguetia flagellaris*(?). ARACEAE: Species indet. MELIACEAE: *Guarea scabra*, *Guarea silvatica*, *Trichilia septentrionalis*. MYRISTICACEAE: *Virola pavonis*.

Stomach-ache
BURSERACEAE: *Protium spruceanum*, *Trattinnickia glaziovii*. MELIACEAE: *Trichilia septentrionalis*. MONIMIACEAE: *Siparuna decipiens*. OLACACEAE: *Ptychopetalum olacoides*.

Cough
ANNONACEAE: *Duguetia flagellaris*. HUMIRIACEAE: *Sacoglottis matogrossensis*. LEGUMINOSAE: *Marmaroxylon racemosum*. RUBIACEAE: *Palicourea quadrifolia*.

Diarrhoea
COMBRETACEAE: *Buchenavia* sp. [278]. LEGUMINOSAE: *Bauhinia guianensis*. MARCGRAVIACEAE: *Norantea guianensis*. RUBIACEAE: *Palicourea quadrifolia*.

Snakebite
CONVOLVULACEAE: *Maripa* sp. [533](?). DILLENIACEAE: *Tetracera amazonica*. LINACEAE: *Roucheria punctata*. PIPERACEAE: *Piper consanguineum*.

Grippe
BIGNONIACEAE: *Tynanthus* sp. [GR2787]. MORACEAE: *Pourouma guianensis*. RUTACEAE: *Citrus aurantifolia*.

Appendix IV continued

Fungal infections (+ mites?)
APOCYNACEAE: *Bonafousia undulata.* MORACEAE: *Naucleopsis amara.* ORCHIDACEAE: *Brassia bidens.*

Gut parasites
LEGUMINOSAE: *Andira* sp. [790]. MONIMIACEAE: *Siparuna decipiens.* MYRISTICACEAE: *Iryanthera juruensis*(?).

Localized aches/pains
EUPHORBIACEAE: *Croton lanjouwensis.* LEGUMINOSAE: *Andira* sp. [790]. VOCHYSIACEAE: *Qualea* sp. [65].

Botfly larvae
EUPHORBIACEAE: *Hevea guianensis.* SAPOTACEAE: *Manilkara bidentata.*

Eye infection
GUTTIFERAE: *Clusia insignis.* MORACEAE: *Ficus guianensis.*

Childbirth
BIGNONIACEAE: *Callichlamys latifolia.*

Headache
LEGUMINOSAE: *Parkia nitida.*

'Liver'
MONIMIACEAE: *Siparuna decipiens.*

Mouth ulcers
LEGUMINOSAE: *Machaerium quinata.*

'Madness'
ANNONACEAE: *Annona ambotay.*

Menstrual disorders
LEGUMINOSAE: *Bauhinia guianensis.*

'Pus'
MORACEAE: *Helianthostylis sprucei.*

Other:

kirita ieptxi
MYRISTICACEAE: *Iryanthera laevis.*

pitxeni kinji' (cure associated with babies)
COMBRETACEAE: *Buchenavia* sp. [278].

pitxeni kipytxiwy (cure associated with babies)
LEGUMINOSAE: *Parkia velutina.* MARCGRAVIACEAE: *Norantea guianensis.* ZINGIBERACEAE: *Renealmia floribunda.*

Unspecified.
EUPHORBIACEAE: *Hyeronima* (?) sp. [616].

INDEX TO SCIENTIFIC (BOTANICAL) NAMES INCLUDED IN THE TEXT